Winter 2006

Editor – Julianne Schultz

Griffith REVIEW is published four times a year by Griffith University in conjunction with ABC Books.

ISSN 1448-2924

Griffith University
University Drive
Meadowbrook Qld 4131
Australia

ABC Books
GPO Box 9994
Sydney NSW 2001
Australia

Telephone (07) 3382 1018
Facsimile (07) 3382 1441
Email griffithreview@griffith.edu.au
Website www.griffith.edu.au/griffithreview

Subscription (4 issues) incl postage and handling
$66.00 within Australia including GST (Recommended Retail Price)
A$108.00 outside Australia
Institutional and bulk rates available on application from the Business Manager at the Griffith University address above.

Letters to the editor should be sent to The Editor, *Griffith REVIEW*, Logan Campus, Griffith University, Meadowbrook, Qld 4131, Australia. Alternatively they can be emailed to griffithreview@griffith.edu.au or by logging on to the website www.griffith.edu.au/griffithreview. The Editor reserves the right to edit letters for publication.

Cover and title page image: Martin Barraud / Source: Gettyimages.com

Sir Samuel Griffith was one of Australia's great early achievers. Twice the premier of Queensland, that state's chief justice and author of its criminal code, he was a great, confident provincial who saw no reason why, as a Queenslander, he could not take a major role in the public life of the emerging nation.

Best known for his pivotal role in drafting agreements that led to Federation and as the new nation's first chief justice, Griffith was also an important reformer and legislator, a practical and cautious man of words – a legal draftsman, a poet and translator of Dante. Griffith, who died in 1920, is now remembered in his namesakes – an electorate, a society, a suburb and a university.

Ninety-six years after he first proposed establishing a university in Brisbane, Griffith University, the city's second, was created. Now Sir Samuel Griffith's commitment to public debate and ideas, his delight in words and art and his attachment to active citizenship have again been recognised by the university that bears his name in the publication of *Griffith REVIEW*.

This quarterly publication is designed to foster and inform public debate, to bridge the expertise of specialists and the curiosity of readers, provide the space to explore issues at greater length with more reflection, and offer the opportunity for established and emerging writers and artists to tease out complexity and propose new ways of thinking and seeing.

Each issue of *Griffith REVIEW* is devoted to a topical theme and includes essays, analysis, reportage, memoir, satire, fiction, poetry, photography and art. The range is deliberately wide. The complexity of contemporary issues cannot be adequately illuminated by one form alone – the who, what, when, where, how and why of events are all important, but the feel for what they mean may better be understood through a novelist's eye, the cacophony of current activities better comprehended with an expert's insight into competing theories or the rhetoric of public debate unpacked by detailed behind-the-scenes reportage. These forms and others will be welcomed in *Griffith REVIEW* as it attempts to capture the spirit of the times and build a bridge between literary, academic and journalistic writing and the reading public.

Like Sir Samuel Griffith, *Griffith REVIEW* will attempt to inform and advance public debate and understanding; it will also be iconoclastic and non-partisan, with a sceptical eye and a pragmatically reforming heart. ■

Contents

MEMOIR

POLICY

POETRY

DEBATE

FICTION

ONLINE

www.griffith.edu.au/griffithreview

Introduction: **Masters of the universe: How nigh's the end?**

Author: **Julianne Schultz**

After listening to the eminent scientist cataloguing the ever-increasing evidence of significant climate change – rising temperatures, rising seas, extreme weather, melting ice – and its potentially apocalyptic consequences, the reporter wanted to know how suddenly this would play out. "It's always a question of what is sudden," Professor John Schellnhuber replied. "The collapse of the western Arctic icesheet may happen over centuries but in a geological time scale, this is very sudden."

Therein lies one of the conundrums of the climate change debate. Sudden depends on how time is measured. For a scientist analysing changing concentrations of greenhouse gases captured over 650,000 years in a three-kilometre-long ice core, a twenty-seven per cent increase in carbon dioxide in a century is sudden, but for a politician locked into a three-year electoral cycle, sudden may be a jump in the polls from one week to the next. To a geologist, sudden may be a change that takes tens of thousands of years, yet play out over millennia for an archaeologist, take a century for a historian or a couple of seasons for a farmer.

In this movie-time world, defined by instant media, most of us expect sudden to be urgent, immediate and pressing. In the movies, a city can be inundated in the time it takes to eat a tub of popcorn; with half an eye on the television, a cyclone can form, denude a rainforest and flood a region between breakfast and dinner. This is the sort of time we have become accustomed to, but it is not particularly helpful in understanding either the causes or potential consequences of climate change. This will become apparent as you read this collection in which very different senses of time inform the perspectives presented.

What is known is that carbon dioxide is now accumulating in the atmosphere more rapidly than at any other time in recorded history and that the consequences of this are likely to be measured in significant changes, rising temperatures, melting ice, rising seas and extreme weather – both hot and cold – within this century. This is not a question of belief, as critics contend, but of evidence that can be tested and measured, drawing on the expertise of scientists from a dozen or more disciplines. They do not all agree, the competition between disciplines can lead to vigorous debates, but the trend line is clear and troubling as the rigorously peer-reviewed Intergovernmental Panel on Climate Change has found.

The deep history of the planet shows that dramatic climate change has occurred in the past and will almost inevitably recur. What is new is the contention that as a result of profligate use of fossil fuels and an ever expanding population we are forcing temperatures to rise more suddenly. The complex questions the scientists have raised can no longer be addressed by science alone.

As befits an information age, more is now known about the world in which we live than could once have even been imagined. It is easy to get lost in the detail, to be uncertain about the science and confused by the rhetoric. It is therefore important to acknowledge that the institutionally conservative great national associations of scientists are united in their diagnosis that the evidence is pointing to dramatic changes in climate that are likely to affect the way we live in future. Most still hold out the hope that something can be done – by adopting renewable energy, modifying behaviour, limiting population growth, taxing unsustainable activities and developing new technologies – but increasing numbers of experts are adopting an even more pessimistic view: it may already be too late.

The sheer complexity of the data that is available about everything, from the air we breathe to the ground we walk on, is astonishing. More is known about the physical environment than has ever been known before. We know, for instance, that three plant or animal species are becoming extinct every hour while 9,000 more human babies are being born, and that the Carboniferous Age that laid down the great stores of fossil fuels 200 million years ago and that powers our world will never be repeated.

The citizens of previous great civilisations may have felt confident that they knew the limits of their world but this did not prevent their demise, as Jared Diamond has so compellingly documented in *Collapse* (Allen Lane, 2005). Changing weather patterns, drought, flood, fire, pestilence and war – the horsemen of the apocalypse – have played a part in the demise of many civilisations. The question is whether we are really such masters of the universe that we can prevent it happening again.

This challenging big-picture view of weather – as an adjunct and ally in the development of human civilisation – is often lost sight of in the daily deluge of new announcements, research, claims, counterclaims and the political posturing that passes for public debate about climate change. Putting global warming in this context provides a stimulating perspective. Much of the knowledge of the evolution of the planet and human life upon it has been acquired, tested and proven relatively recently. The very notion and capacity to measure the gases in the atmosphere, or trapped in the ice or in the remains of a human being tens of thousands of years old, is science that has evolved over less than a century – almost suddenly.

Although more is now known, questions remain about whether that knowledge can be meaningfully applied in a way that draws from the lessons of history. If we were able to take and apply this knowledge to ensure the survival of the planet and civilisation as we know it, it would be an unprecedented feat of imagination and initiative. Such a project has never been attempted before; its portents are not good. Even if it succeeded, it could still be set to naught by natural systems beyond our control: tsunamis, volcanoes, asteroids and worse, as Murray Sayle notes in his remarkable essay, which ranges across time and space and traces the origins of our enduring fear of an apocalyptic end. He points to the problem that rests with our social and economic organisation, and notes that while the outcomes are unknowable, a solution of sorts may in part grow from sharper recognition that we all have only one earth, our home, Emoh Ruo.

Sayle finds in the rise and fall of the Dutch Republic – which in the sixteenth and seventeenth centuries created a system of globalised capitalism and *en passant* discovered and named Tasmania – parallels to our time. It is ironic that the people of the Netherlands should provide this inspiration, as theirs is a country, with a quarter of the landmass below sea level, among the most vulnerable to the rising sea levels that are expected to accompany global climate change.

The long-term perspective that informs Sayle's essay is both refreshing and challenging. It puts contemporary science into a historical and economic context in a way that has not been attempted before, and reveals with searing clarity the underlying problem, an issue occasionally referred to as the Cinderella of climate change. The quandary that goes to the heart of sustainable life on Earth is the sheer number of people living here: from one and half billion people a little more than a century ago, to six and a half billion human beings this month and nine billion in another fifty years. The exponential increase in global population and the resources that are required to sustain it continues even as the birthrate stalls in many developed countries.

The International Energy Agency predicted in 2004 that as a result of industrial and population growth the global demand for energy will increase by sixty percent by 2030. Eighty-five percent of that growth will be derived from fossil fuels, especially coal in China and India. Despite this enormous growth the IEA predicts that by 2030 the number of people with access to electricity will have fallen only slightly from 1.6 billion to 1.4 billion and 2.6 billion people will still be using traditional biomass for cooking and heating.

For a world with finite resources, and with an underlying, if periodically challenged, respect for human life, this presents almost unimaginable dilemmas and highlights the urgency and enormity of the problem. Many are confident that human ingenuity will solve the problem that by switching to renewable energy, by fostering the economic growth that means women have fewer babies, by learning to live in a sustainable way, life on Earth may continue to be viable. Others are less sanguine. Political leaders of all hues and geographic origin are declaring that global warming is the greatest challenge confronting the globe. Some even warn of a future beleaguered by wars over resources, with fighting not just over oil, but water, land, food – like something out of a dire, futuristic movie.

The seriousness with which climate change is being taken by the rest of the world comes as something of a shock to Australians, insulated as we are by land and distance and a political discourse that is sceptical about science but trusts technological solutions. The data that shows Australians are among the most profligate users of fossil fuels – as befits the world's largest exporter of coal – is one of those abstractions that is easy to accept but hard to comprehend. Life in this country is not much touched by poisoned air and water, or the malodorous overcrowding typical in so many other places. Australians have always recognised the power of weather and the reality of climate change is beginning to gain popular currency, due in no small measure to news of extreme weather events, which may not themselves be a direct consequence of global warming, but contribute to a sense of foreboding. Droughts, El Niño, water shortages, shrinking snow, cyclones and searing temperatures remind us with increasing regularity that there are challenges in even this vast remote and ancient land.

For most, life is good, thanks in no small measure to the money generated from the sale of the prodigious quantities of fossil fuels and other resources that China needs for its industrialisation, and the ability to buy the cheap products manufactured in the plants fuelled by these resources – what the governor of the Reserve Bank called the China Factor. This takes the sting of urgency out of the debate here. Nothing is likely to happen suddenly, Australians are comfortable with complacency. Meanwhile many countries – including to a very limited degree even China and India – are wondering about they might continue to grow with less environmental impact. Australia's wealth of fossil fuels neuters any will or incentive to capitalise on renewable energy for domestic use or global benefit.

Most Australians care deeply about the local environment – time and again it registers as the most pressing matter. Yet the consequences of global warming are likely to be profound, both personally and economically, at home and beyond. The CSIRO predicts that Australia will become more arid, that the average temperature is likely to rise by at least one to six degrees or more over the next sixty-five years, sea levels will rise, El Niño will become even more frequent, the Great Barrier Reef will continue to deteriorate, fierce fires will become commonplace and tropical cyclones more frequent and fiercer. Accompanying this will be habitat destruction and species extinction. The regional challenges will be even more serious.

It is possible that the economic and security consequences of global warming may force the issue onto the political agenda. As major insurance companies factor the costs of climate change and a market in trading carbon credits evolves, Australian business will be reluctant to miss out. There are plenty of advocates of the opportunities that could come from recognising the challenge and seeking to adapt to it. Even President Bush, his own wealth derived from oil, has cautioned his energy-hungry nation to seek to break its addiction to oil, but failed to propose any sanctions to achieve it.

Similarly, with more than a hundred million people living within a metre of mean sea level, many within this region, predicted rises in sea levels are likely to create another category of refugee – those fleeing environmentally unsustainable homelands. The future is likely to be volatile – the weather in one form or another may again determine our survival.

None of this is likely to happen suddenly in the popular understanding of the word, but the weight of evidence suggests that the march of climate change is well under way and this is the fleeting moment in which it must be addressed. It may be too late, the capacity for us to understand the information that is available and act wisely on it may be beyond our capacity for imagination, organisation and action – or the costs may be deemed too great and the benefits for as yet unborn generations too slight. But the burden and responsibility of knowing something is up rests heavily and uncomfortably on our shoulders. ■

Letters to the editor and extended contributors details available online at www.griffith.edu.au/griffithreview

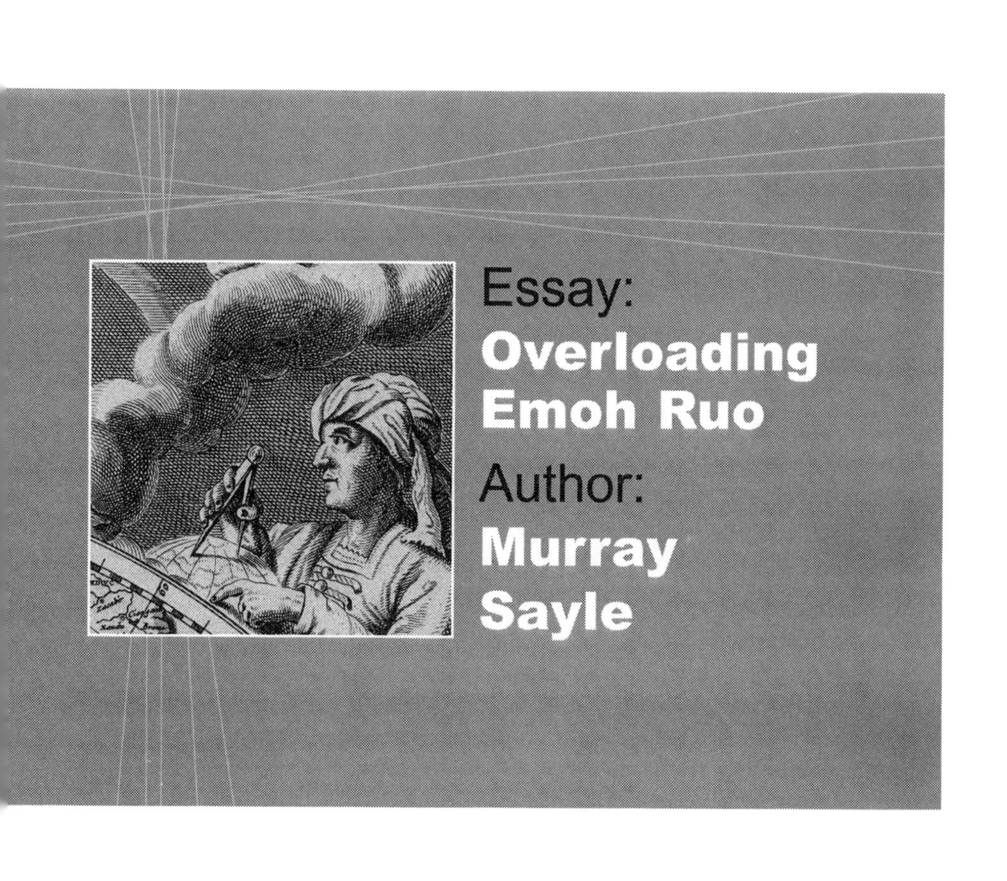

Essay:

Overloading Emoh Ruo

Author:

Murray Sayle

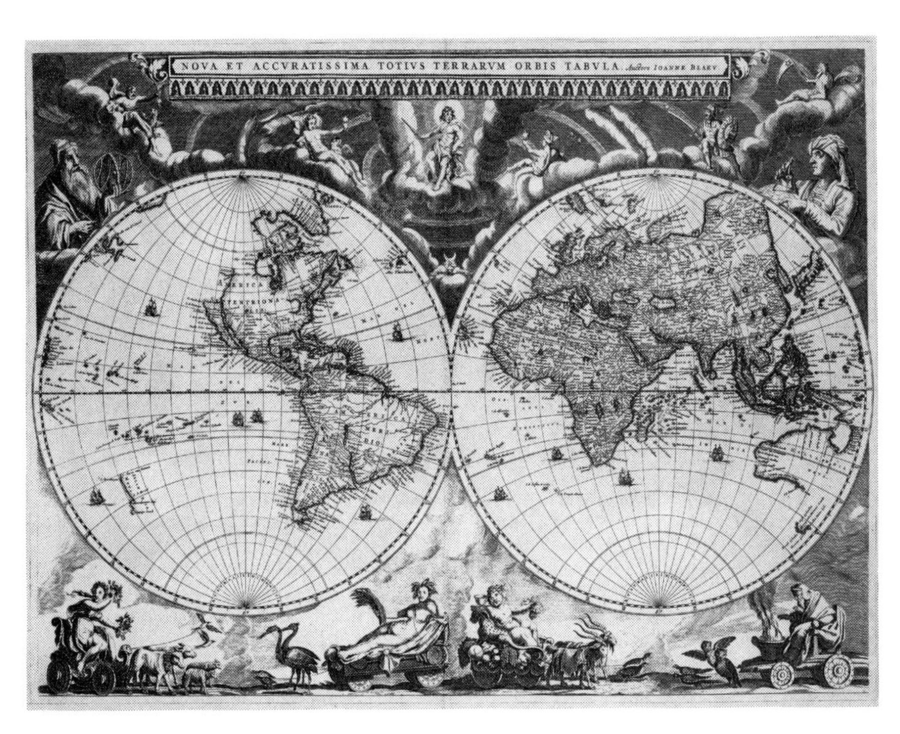

Image: Joan Blaeu / Nova et accuratissima totius terrarum orbis tabula [New and very accurate chart of the world], 1664? / handcoloured engraving / Courtesy of the Dixson Library, State Library of NSW.

Overloading Emoh Ruo: the rise and rise of hydrocarbon civilisation

No one discovered global warming, although it has been going on for most of the past 12,000 years. The suspicion that human activity is altering the world's climate forced itself in slow stages on disbelieving scientists, as it is now reaching a wider but only half-persuaded public. The science is complex, the factual forecasts tentative at best. One simple truth is, however, now established beyond any doubt, one of those rare insights that will change the way we think as long as our species lasts. We have only one single human household, the much-abused planet under our feet. We are hopeless at managing it, a dysfunctional family riven by jealousy, selfishness and self-imposed ignorance. We show no real sign of changing our ways. If we go on like this, the outlook we refuse to face up to is grim.

Cape Grim Baseline Air Pollution Station is a modest collection of prefab huts atop a blackened tooth of sandstone that juts from the north-west corner of Tasmania into the Southern Ocean at latitude forty degrees forty-one minutes south. This is uncomfortably inside the Roaring Forties, the winds that blow clear around the globe with only a short passage over the distant wilds of Patagonia, raising swells that break with few pauses on two saw-toothed islets at Grim's foot, ninety-four metres below. Above the huts, which house a rotating team of nine scientists and their instruments, soars a slender steel tower, with air intakes sampling the winds at 104 metres and 164 metres above the sea.

Grim was given its peculiarly prophetic name by Matthew Flinders, RN, then aged twenty-four, who set out from Sydney in the southern summer of 1798 with surgeon George Bass, twenty-six, in the seventeen-ton sloop *Norfolk* to test whether Van Diemen's Land, as it was then known, was part of New South Wales or, as many suspected, an island in its own right. The two youthful shipmates sailed (and named) Bass Strait from east to west and, threading their way through the Hunter Isles at the strait's western mouth, kept careful watch for the unknown coast to bear away to the south-east, conclusive proof of their island thesis. Sure enough it did. Introducing his *Voyage to Terra Australis* (which, incidentally, gave this country its name and has caused so much grief to Austrian postmasters), Flinders wrote: "The north-west cape of Van Diemen's Land, or island, as it might now be termed, is a steep black head which, from its appearance, I call Cape Grim."

The windy cape stemmed the Roaring Forties in lonely isolation until 1976, a year in which, with the Vietnam War ended and the Middle East unusually quiet, thoughts turned to meteorology, the science of weather, an area in which even the most self-centred of nations can see advantages in exchanging information. While military men keep a covert eye on the world climate, they like to show themselves co-operative in a good cause and, as children everywhere have discovered, what sounds more seductive than "You show me yours and I'll show you mine"?

It was a time of bizarre weather events. In the early 1970s, India's life-giving monsoon failed and there were poor harvests in Africa and the then Soviet Union. Cities worldwide were getting grimier, with black-rimmed collars and high-rent urban vistas shrouded in smog – not the Great 1953 Killer Smog, caused by burning coal in a million grates, which paralysed London for three days and killed thousands, but something more complex and mysterious. In 1975, the World Meteorological Organisation, an offshoot of the United Nations, called for member states to finance a worldwide chain of twenty-two reporting stations in out-of-the way places to investigate these odd events. No one, for once, vetoed the innocent-sounding proposal. The decision gave Cape Grim a new global significance.

There was a precedent, of sorts. In 1957, another quiet year internationally, the United Nations proclaimed an International Geophysical Year, partly inspired by the launch of the first man-made satellite, *Sputnik*. That year, an American scientist-hero, Charles Keeling, set up an air-sampling station atop Moana Loa, a 4,169-metre peak in the Hawaiian Islands, surrounded by thousands of kilometres of pristine ocean air. Keeling nursed a suspicion that carbon dioxide, CO_2, was somehow changing the climate. A genius at scrounging funds, Keeling almost ran out of money several times, but kept on refining his methods and measuring the air.

Sure enough, his findings showed a steady, steplike climb of airborne CO_2 between 1957 and 1975, but the quantities were still minute. Most of us know that the air we breathe through our air-conditioners is roughly four-fifths nitrogen and its derivatives, now rising; a fifth is oxygen, falling; and less than one per cent other gases and odds and ends, some nasty. Keeling's air samples showed an increase of CO_2 from 315 to 330 parts per million of air in almost twenty years. But Moana Loa is an active volcano, volcanoes spew CO_2, among other gases, and maybe this was skewing Keeling's findings.

All that fuss, his critics scoffed, about a wisp of harmless gas. It is true that the danger, if there is one, is far from obvious. We all breathe out about a kilo of CO_2 every day of our lives. It makes the bubbles in champagne, the froth on beer. It forms the swirling clouds of vapour through which ageing rock stars make their entries. Shopkeepers still pop a pellet or two of dry ice, the solid form of CO_2, into parcels of frozen food, and children find they bubble briskly underwater. Could this fun gas really end life as we know it? Keeling visited the Antarctic in 1957, not exactly crowded with human activity, and found that CO_2 levels in the frosty air climbed in a single season. The pattern he detected at Moana Loa was repeated in the pristine Antarctic.

Australian weather scientists had long seen the possibilities of Flinders' sinister Cape Grim for a "baseline" station. It was an ideal spot to measure the atmosphere's minimum level of CO_2 and other suspect gases, before human or natural activity, nearby or distant, could add to it. A quirk of history had kept Grim all but untouched by human boots. In 1976, the Cape was still part of a grant of 22,000 hectares (220 square kilometres) made in 1825 to the London-based Van Diemen's Land Company (VDLC), intended to exploit the raising of fine wool in the then remote Tasmanian north-west, and thus called Woolnorth. The station still flourishes and still belongs to the VDLC, the only surviving Royal Charter Company, which also has holdings in New Zealand. The last volcanic activity in the region ended some twelve million years ago.

Woolnorth, originally staffed by indentured labourers from England with their wives and children, and ticket-of-leave convicts, got off to a brutal start, with records of "depredations" by the local natives – a sad glimpse of Tasmanian Aborigines, soon to disappear from human history – and of unauthorised, violent reprisals by the company's well-armed "servants". Woolnorth only began to prosper in the 1850s when gold was found in nearby Victoria and the VDLC developed Woolnorth as a supplier of beef, mutton and potatoes for the diggings; it has continue to prosper as a supplier of fine wool.

Bare and roadless, Cape Grim was unsuited for these profitable lines and the Australian Bureau of Meteorology (BOM) and the Commonwealth Scientific and Industrial Research Organisation (CSIRO) acquired title to the tip of the cape with a buffer strip between it and Woolnorth in 1957. It was to become one of the World Meteorological Organisation's baseline monitoring stations. The first air samples were collected in April 1976, the brief being to analyse the four so-called greenhouse gases: carbon dioxide, methane, nitrous oxide and sulphate aerosols, which are co-emitted with airborne soot.

Carbon dioxide lingers for 150 years or more in the atmosphere and so has attracted the most nervous column inches. Methane, otherwise known as swamp gas, has more drastic affects on the climate but disperses after a decade or so. Nitrous oxide is a compound of oxygen and nitrogen, the main ingredients of air, and eventually breaks down. Sulphate aerosols, which are not gases, are better known to us as the smog that dims our cities. All are natural products and are present in all air samples. Methane bubbles up from vegetation decaying in swamps, or in the long swamp-like intestines of grass-eating animals or, in layperson's language, from farting cows. Nitrous oxide is produced when lightning heats air. Some sulphate aerosols come from volcanoes, some from bushfires.

Sure enough, ice cores record the four substances fluctuating in no discernible pattern over many millennia. Then, beginning about the end of the eighteenth century, the four begin what looks like a concerted climb. In the documents issued by the fifth session of the Intergovernmental Panel on Climate Change (IPCC) in Nairobi in April 2001 – the smoking gun of global warming, if there is one – the concentration of the four substances in the cores faithfully track each other until, as Grim and the other baseline stations take over after 1976, they approach the vertical. No one has come up with a better explanation than the obvious: the graphs show the consequences of human activity, the oncoming industrial, fossil-fuel-fired revolution, the still incomplete triumph of our own beloved hydrocarbon civilisation.

The views from Cape Grim are awe-inspiring – to the north, Bass Strait, to the west, the empty ocean, and, curving away to the south-east, Valley Bay, which convinced Bass and Flinders that Van Dieman's Land was indeed an island. In the distance is the wind farm operated by Hydro Tasmania, whose thirty-seven giant rotors power the whole operation and feed surplus electricity into the Tasmanian grid. But no one lives at Grim on a permanent basis. The offices and living quarters of the scientists and technicians are sixty kilometres away, in the nearest town, Smithton (pop. 3,495), a cheerful place named, like much else in these parts, for a long-dead director of the VDLC. For its size, Smithton has every mod con, with two pubs, three supermarkets, two ATMs, a hospital and a swimming pool filled from the Duck River on which the town stands.

The analyses of the air samples collected at Grim are automatically sent by the web to Smithton, to the CSIRO in Melbourne, the Woods Hole Oceanographic Institute in Massachusetts, United States, and other interested parties. But Grim's remit is to study not simply the baseline of CO_2 in the atmosphere, but some other so-called "greenhouse gases" as well: methane (CH_4), nitrous oxide (N_2O) and a group of human-made gases incorporating fluorine or chlorine that do not occur in nature and for which there is therefore a simple, if heroic, remedy – stop making them. Constant refinements in detecting these gases require frequent visits to Grim by the scientists and technicians, and occasional overnight stays.

Alas, the term "greenhouse gas" is a metaphor which, like many such, started off as a brave attempt to make a complex idea clear and wound up befogging understanding even more and opening the way to charges of deceit and even fraud. To make things worse, it omits the most potent "greenhouse gas" of all – H_2O, plain, ordinary, tap, river or seawater in vapour form. The truth is, your common or garden greenhouse has no greenhouse gases. It works by letting sunshine through its clear, glass or plastic walls. Homonuclear, diatomic molecules – a mouthful meaning the simple, healthy gases like oxygen (O_2) and nitrogen (N_2) of air – do not absorb any energy from sunlight, which shines through on to the earth floor of the greenhouse, warming it and the air just above it. Warm air rises, and so the air in the greenhouse warms up, trapped by the clear walls. It is said that greenhouse operators in cold climates like the United Kingdom have been known to add CO_2 as well, but this is rumoured to give their tomatoes a woody taste and so is best avoided. Our Earth has no walls and is not a giant greenhouse – any rising, warm air is dispersed by the winds and mixed by turbulence. The principle involved is different and the other so-called greenhouse gases have nothing to do with greenhouses, real or metaphorical; sulphate aerosols are not even a gas. Still, "greenhouse" is easier to remember than homonuclear or diatomic, so we are stuck with the term, and had best understand it.

The first thinker to have brought CO_2 into the speculation on climate change was a Swedish chemist named Svante Arrhenius, in 1896. He was then engaged in one of the hottest intellectual searches of the day for what had caused the ice ages. Until the middle of the nineteenth century, most people who thought about such matters were comfortable with the calculation of Archbishop James Ussher of Armagh in Northern Ireland who, by totting up the begats in the Old Testament, had worked out that the world was created on Saturday, October 22, 4004BC, at 6pm, so as to make Sunday, October 23, the first full working day of creation.

As mankind's last century of peace and hope, the nineteenth, progressed, evidence began to accumulate that the glaciers of Europe were melting, leaving characteristic heaps of ice-smoothed boulders behind. Such ice-gouged valleys and telltale piles of stones are to be found all over northern Europe, Asia and North America, revealing that as recently as 12,000 years ago, the areas where London, Stockholm, New York and many other cities now stand were buried under enormous sheets of ice. The oceans were then at an all-time low and early men walked barefoot from Asia to Japan and America and over the Bass Strait to what is now the island of Tasmania. These ice ages had happened not once but many times in the distant past. In one, the Earth had been a giant snowball, entirely wreathed in ice, which must have been a pretty sight. Strictly speaking, what we have had lately are mini-versions of the great snowball, more properly called glacial and interglacial periods, though we have come to call them all ice ages. "What had caused them?" Victorian-era scientists wondered. Could they happen again?

Indeed they could, argued Arrhenius, and from a chilly Swedish point of view this might be a good thing, just as some Russians now look forward to the first vintage of Moscow grapes. That weather is changeable we all know, and from our own experience we can see some of the reasons. Liquids like champagne hold more dissolved gases the colder they are, and these are driven off as bubbles as they warm up. Cold water contracts and sinks, hot expands and rises, which is why we stir a baby's bathwater before putting the little cherub in. The atmosphere is the other way round. Hot air holds more dissolved gas, particularly water vapour, cold air less, so that our days are either hot and muggy or cold and dry, unless it is actually raining. These contradictory tendencies interact in ways that defy short-term prediction, which is why meteorologists keep umbrellas in their offices. Yet, in a general way, the weather follows a pattern – we put our summer clothes away for the winter, and vice-versa.

Arrhenius thought about other phenomena that are hard or impossible to predict in the short term and yet almost entirely predictable in the long run, and came up with volcanic eruptions. He knew these spew enormous quantities of gases, including CO_2. This would raise air temperatures somewhat, which would increase humidity, which would increase warming, which would further increase humidity, and so on – a small input leading to a bigger output, what engineers call positive feedback. Conversely, argued Arhenius, if volcanic eruptions should slow or cease, the reverse process could see the CO_2 already aloft being reabsorbed into the soil or seawater, vapour in the air would fall and, in the fullness of time, a new ice age would be upon us.

This was a brave try, and on the right track. No one else had tried to factor humidity into the long-term weather or climatic equation. Arrhenius toiled long and late on his theory, laboriously calculating by hand estimates of humidity for every degree of latitude in the northern hemisphere and consulting such old records as he could find. His conclusions left the growing world scientific community distinctly unimpressed and the loyal followers of Archbishop Ussher jubilant. One experimenter tried passing infra-red radiation (the most warming kind) through a column filled with CO_2, about the same quantity as was then found in the atmosphere. There was virtually no change when the quantity of CO_2 was doubled. No one then realised that CO_2 absorbs radiation only in certain narrow bands of the spectrum, and that a mere trace of the gas produces bands as opaque as planks, so that adding more makes little or no difference. Writers hoping to humanise the dour, driven Arrhenius have attributed his burning of the midnight electricity to an attempt to forget the pain of a divorce and a bitter custody battle over his son. From this distance, he seemed more like a man seized by an idea that would give him no peace. In any event, in 1903 he became the first Swede to win the new Nobel Prize, not for his CO_2 theory but for his discovery of ionisation, one of the most powerful insights of physical chemistry. His climatic theory became a neglected curiosity and if it influenced Keeling at Moana Loa, this awaits documentation.

It was to be another seventy years before evidence came to light supporting both these theorists, and then from a totally unexpected direction, deep below the Earth's surface characteristics.

The last ice age, we may almost recall, is only about 12,000 years behind us. The period since then, which ecologists call "the long summer", has encompassed the entire history of our agriculture, cities, writing, science, printing and the publication of university reviews, not to mention buying new cars and watching TV game shows. At the depth of that chilly but creative time, the CO_2 in our bracing atmosphere was down to 180 parts per million of air, about as low as it has gone since there has been life on Earth. We know all this with such precision because the ice ages left two enormous iceblocks, which have never melted since the big freeze began. One covers Greenland, the other most of the Antarctic continent. It is noticeable that they are both on dry land standing above sea level. The sea cannot rise high enough to melt them. As they are snow-covered, and therefore white, they reflect almost all the sun's heat back into the sky, as the whole Earth did in the years of the great global snowball. The Arctic is, in contrast, an ocean with a thin skin of ice. Submarines have not only cruised under it but cheekily surfaced in one of the clear patches of water that have been appearing lately near the North Pole. Nature's two slumbering ice lollies guarded their secrets until the 1970s, when scientists began drilling into them, starting in Greenland, and using techniques developed by geologists to search for oil and other saleable minerals. Air carries moisture, and changes in air pressure cause precipitation in the form of snow on the ice. The result is to leave layers of snow-turned-into-ice, rather like the annual rings of a tree. Trapped in the ice are tiny bubbles of air brought down with the snow.

By a wonder of modern science, it is possible to analyse this air by heating it to luminescence and observing bands in the resulting spectrum of light. The most recent of these borings, conducted by the European Union at its Concordia site in Antarctica, reached a record depth of 3.2 kilometres. Similar drilling has taken place in deep-ocean sediments and the floors of ancient lakes. Taken with the geological record, they tell us that CO_2 levels in the air and sea levels around our world have risen and fallen in rough synchrony for the past 650,000 years, a time that has seen six full ice-age cycles plus a half, the one we're in now – with the data frozen unarguably in ice.

This gives us a glacial/interglacial period of roughly one every 100,000 years. We cannot date the rising and falling seas with any such precision, of course. But judging by the CO_2 levels, what we see is what we would expect from a number of natural processes sometimes co-operating and sometimes in conflict, but tending in the same general direction – not a steady beat like a healthy human pulse but jagged lines with a regularly irregular pattern, a peak, or sometimes several, in airborne CO_2 levels, a steady decline with many short-term reversals we might call peaklets, a low of both CO_2 and sea levels, meaning maximum ice, and then a steep, almost cliff-like climb towards a new high CO_2, higher seas, lower ice peak. All of this was happening in geologic time, so that an individual Mr Hunter or Ms Gatherer might or might not notice a change from year to year like the changes we are, or think we are, seeing now.

The last set of peaklets, covering something like 100,000 relatively icy years, is of particular interest to us as they cover the entire history of anatomically modern man, *Homo sapiens sapiens*, the "wisest of the wise", as we call ourselves (how our successors, if we have any, will smile at our conceit) – the big-headed, weak-jawed, sway-backed creatures you can see gathered around the office water cooler or, indeed, crouched at this word processor. In that short space of time, *Homo sapiens sapiens* perfected the use of symbolic language, walked out of Africa across the Suez land bridge, interbred with or wiped out our predecessors *Homo erectus, Homo neanderthalis* and company and then – at least the most adventurous of us – made it over open-sea crossings to far-off Australia-to-be, arriving some 50,000 years ago. Yet the legends of the Middle Eastern peoples, full of floods, famines, whirlwinds and even the odd whale, tell nothing of advancing and retreating sheets of ice. The only reasonable explanation – another speciality, we believe, of *Homo sapiens sapiens* – is that Archbishop Ussher's biblical sources must have arrived from the other direction, after that human battle for survival had been won. In that case, give or take a few begats, his chronology may well be correct, his informants mistaking their own arrival for the beginning of things, as every new civilisation or generation does. Equipped with a portable religion whose sacred object was a book, the ancient Hebrews preserved their memories, which we still consult.

We know that the ice eventually melted, the seas rose, and Stockholm, London and New York became tidal ports. In the fullness of time, Sydney residents fought for harbour frontages to moor their yachts. From its seldom-exceeded low of about 280 parts of CO_2 per million of air, the ice cores tell us that CO_2 began to rise about 15,000BC – as indeed it was due, even overdue, to do.

Here we need to shorten our focus and get out a metaphorical magnifying glass. Some time around 1750, the CO_2 level started to climb into the stratosphere, unsuspected by anyone on Earth. By 1900, it was around 300 parts per million of air, close to the cores' historic high. From there, it turned into what statisticians call a "hockey stick", a gentle curve developing a sharp upward hook on the end. When Keeling began operations on Moana Loa in 1956, it was 315 parts per million. By 1976, when Grim began sampling the Roaring Forties, it was 325.

When I checked for this article in early 2006, Grim was reporting 376 parts per million, pushing on for 380. It was already twenty-seven per cent higher than the peak recorded by the deep ice and ocean cores over the past 650,000 years. A conclusion seems obvious and may indeed be correct. The years 1750 to the present cover, very roughly, the Industrial Revolution, first in Europe, then in Asia. The Industrial Revolution involved, at least in its early stages, the burning of ever more fossil fuels. Fossil-fuel burning gives off CO_2. Therefore, to prevent or slow the warming of the climate, the melting of more ice and the rising of the oceans (they still have fifteen metres to go to reach recent historic highs) we should either burn less fossil fuels or find some way of keeping the resulting CO_2 out of the atmosphere. All the proposals I have seen are variations, earnest or otherwise, on this simplistic but unworkable theme.

They ignore another statistic, seldom mentioned, tidings even grimmer than Grim. About now, Earth's human population is passing a stupendous milestone. There are now 6,500 million of us, six and a half billion for the zero-averse. These figures come from the World Census of the US Census Bureau, which gives March 1, 2006, as the date of destiny, and the UN Population Office, which uses slightly different criteria and makes it October 1, 2006. As of now, 249 are born for every ten thousand alive and only 108 die, a net increase of six million a month, almost four Australias a year. At this rate we may hit seven billion before 2050.

These statistics go far to explain why Arrhenius was scoffed at or ignored. In 1896, the world's population was between a billion and a billion and a half. Most of these were peasants – poor farmers growing for subsistence. We know that for every molecule of carbon in the atmosphere there are fifty in the oceans. The CO_2 of 1896 probably wound up in the seas, or in growing trees, suggesting a possible upper limit of sustainable human population. But the billion and a half of Arrhenius's day, if we were that numerous, is now six and a half, a fourfold increase. And with the worldwide spread of fossil-fuelled, or hydrocarbon, civilisation, we now average four times the CO_2 output per head of a century ago. Sixteen times the load has overwhelmed whatever kept the system in rough equilibrium, at least short term, and destroyed the comfortable "balance of nature" the nineteenth-century sages talked about. We have deeper questions. What is driving the surge of CO_2 into the air, and with it global warming? The spread of hydrocarbon civilisation. But what drives that spread, and is it sustainable? And if it is not, what happens next, and when? For answers, at least provisional, we need to follow Bass and Flinders on their memorable voyage of discovery.

The peoples of the Netherlands, whom we inaccurately call the Dutch, have been battling climate change and rising sea levels since the end of the last ice age, some ten thousand years ago, and can claim more success than anyone else – using a technique that, the way we are going, we may all soon need. With its 16.5 million people crowded into 41,864 square kilometres, the present-day Netherlands is among the most densely populated of countries, especially considering that a fifth of its area is water – lakes, rivers and canals – and another fifth is reclaimed land, mostly below sea level. That this small gem of human diligence and ingenuity was once a world naval and trading power is one of history's wonders, highly relevant to our own climatically twitchy times.

When the last icesheet covering the North Sea melted, it uncovered the shared delta of two major rivers, the Rhine leading into the heart of Germany and the Maas (Meuse) doing the same for eastern France. The result was an intricate maze of thin, sandy soils left by previous glacial action with stretches of fertile alluvial mud brought down by the rivers. The mouths of navigable rivers, where sea meets land, have often become the sites of great cities, the risk of flood and plunder offset by the attractions of trade, as Shanghai and New Orleans attest. As the North Sea ice retreated, leaving an offshore barrier of sand dunes, adventurous farmers and fisherfolk entered the wet wilderness from higher ground, clustering around laboriously constructed mounds, *terpen*, where they took refuge from river floods and storm surges.

The Roman historian Tacitus reported that the swamps and sandbanks in the delta of the Rhine were inhabited by fierce tribes of Germanic origin he called Batavians, who lived mostly on fish and seabirds' eggs. Wisely, the Romans did not wade into them but accepted them as external allies, and occasionally as mercenaries, notably for their garrison in Britain. Around AD1000, the northern hemisphere was in the midst of the warm period in which 200 Norse (Viking) families from Iceland settled on the west coast of Greenland and others established a base at L'Anse aux Meadows on the northern tip of Newfoundland. They may also have settled somewhere in North America, where they found grapes, and so called the region Vinland.

The melting glaciers, however, raised sea levels and forced a battle for survival on the inhabitants of the Rhine/Rhone delta. As the waters rose, floods and storm surges compelled the building of dikes around some 2,000 small delta towns, at first by the townspeople themselves, then by a mobile force of engineers and labourers paid by the town drainage boards, who levied the citizens. The delta-dwellers had long experimented with wind-driven pumps that had to be turned into the wind and so were limited in power. Early in the 1500s, some unknown Dutch genius developed the *bovenkruier* style of windmill, where only the cap turned to keep the sails headed into the wind. Not only was the characteristic Dutch landscape taking shape, but the Hollanders had found a new source of energy and a promising line of exports. A *moliergang*, a line of self-tending windmills (an ancestor of wind farms), could be set to pump out a shallow lake, uncovering the rotting vegetation of many centuries – an early stage of coal – which could be cut, dried and sold as peat, both useful for home heating and as fuel to distil gin, which could be drunk locally, in vast quantities, and shipped to a thirsty world.

Then, around AD1550, the climate changed again and Europe entered the Little Ice Age, which lasted on and off until the 1850s. The glaciers grew again, Londoners skated on the Thames, the North Sea fell and with it the constant nagging threat to the delta-dwellers, who had, by courage, stubbornness and hard work, laid the economic foundations for their independence and an improbable rise to world power.

Religion, that ennobler of individuals and potent poisoner of human affairs, was central to the story. Through complicated dynastic manoeuvres, the self-appointed sword arm of Catholicism, Spain, laid claim to all seventeen provinces of the Lowlands. Like the Romans, the Spaniards avoided a physical presence and ruled through Dutch-born deputies who bore the hereditary title *stathouder*, a kind of lieutenant-governor. Controlling some of the busiest waterways of Europe, the Netherlands was soon emotionally awash in the passions unloosed by the Reformation, finding that its rejection of distant, dry-shod external authority and stress on individual conscience was an apt expression of the "Batavian spirit", in the phrase of the Anglo-Dutch historian Simon Schama, the joyless Calvinism thundering from Geneva being especially attractive to the smaller, poorer, stiffly self-righteous towns, the Dutch Bible Belt. When Spain's sea power went down with the Armada in 1588, the seven provinces of the Northern Netherlands seized the chance to declare themselves an independent republic, the States General, with its seat of government at Ten Haag (The Hague), where it still is.

Religious differences did not, however, prevent the hard-headed Hollanders from discovering and profiting from a nice little earner. Some time during the Little Ice Age, huge schools of herring migrated for unknown reasons from the Baltic to the North Sea, leaving hungry herring lovers behind. About the same time, the tightening of Roman Catholic discipline with the Counter-Reformation increased the demand for pickled herring to brighten meatless Fridays in the European hinterlands. The Dutch were well placed to satisfy both wants, limited only by the profitless time spent sailing to the fishing grounds and returning to salt down the catch ashore. Even before independence they had developed the herring buss, a factory ship with sixteen to eighteen crew who gutted and salted the herring at sea in six-week spells. These seagoing behemoths were far beyond the resources of individual fishermen. They were financed by the herring merchants of Amsterdam, who took up shares on the long-established model of the municipal drainage boards. Another key element of our modern world was taking shape, all unsuspected, beside the North Sea.

The defeat of the Spanish Armada unlocked the world for Dutch merchants and their ships. Where once they had been small-scale fisherfolk and farmers, they expanded to a near-monopoly of the grain trade from Prussia and Poland to the Mediterranean, of Swedish copper and Stockholm tar, of wool from England and, beginning with a few accidental strandings on their own beaches, of sperm-whale oil for the lamps of Europe, a trade that eventually took them from overhunted Spitzbergen and Novaya Zemlya as far as the Bering and Davis Straits, to Baffin Island and the White Sea. At one time, an entire quarter of Amsterdam, known as the *stinkerijen*, was devoted to boiling down whale blubber and warehousing oil. "The merchants of Amsterdam held their noses and patted their purses," records Schama.

These successes pointed to the biggest prize of all, the spices of East India, in hot demand all over meat-eating Europe and, up to that point, a monopoly of the Portuguese, not long brought under Spanish rule. At first, the Dutch merchants, failing to find a north-east passage to the Far East, tried single-ship voyages to the Spice Islands (now collectively called Indonesia) via Africa and the Indian Ocean. The natives were ready enough to trade and the Portuguese were somnolent but the risks were enormous and, back home, the traders competed with each other, driving down prices.

The loss of lives and worse, of investments, drove the normally standoffish Dutch cities into the kind of arrangement Australians have learnt to call a single-desk strategy or, more bluntly, a monopoly cartel. In 1602, Dutch officials, with much haggling and arm-twisting, organised the merchants of the Netherlands into the United East Indies Company (Verenigde Oostindische Compagnie, VOC). The fastest out-and-back voyage to the Indies took at least twenty months – the equivalent, these days, of doing business by rocket ship with the nearest star in our galaxy – so the VOC was given enormous powers, to build forts, raise armies and make treaties with Asian potentates, almost those of an independent government, which, for all practical purposes, it became. The financial arrangements were equally innovative. Anyone with spare cash could invest in the VOC, but the company was controlled by seventeen directors, in proportion to the trading cities they represented (Amsterdam had half) with the seventeenth as chairman and tie breaker. The funds were invested, not in individual voyages, but in the company itself. The obligations of shareholders were limited to their investments. Shares could not be cashed in, but could be sold at the local commodity market, the Beurs, which thus became the world's first continuously operating stock exchange. Here we see, in embryo, the financial system that now runs our overcrowded world.

At the initial stock offering, 1,200 investors contributed 6,425,588 guilders, an enormous sum for the time, which by VOC rules was a permanent block of capital never to be liquidated or distributed. The first voyages – which took most of the company's capital – were billed as a huge success, with profits of 125 per cent on fast-moving cargoes of pepper, cloves, nutmeg and mace. Some critics pointed out that much of this profit was distributed in kind to the shareholders, at prices set by the VOC, but as the important ones were themselves merchants, this argument could as convincingly show the advantages that come with a well-run monopoly. In any event, the new system showed a well-marked, still powerful feature – to attract more investors, and satisfy the existing ones, the enterprise had to have, or seem to have, rapid continuous growth, the only inducement to offset the risk being the investors' expectation of getting more out than they had put in. The VOC duly expanded at high speed.

The visionary who saw our own future beckoning was the company's first Governor-General, Jan Pieterszoon Coen. Realising the need for a fortified base and entrepôt actually in the Spice Islands, in 1619 he set up shop in – where else? – Batavia, now Jakarta, whose old part of town still has the neat look of a Dutch trading port. To get hold of spice at low prices, Coen at first experimented with slavery, a business with which the Dutch had a long experience in the West Indies, but soon found that forced trade, in which VOC men set the prices at the point of a blunderbuss, was far more cost-effective. In 1624, the company took advantage of a civil war ravaging China to set up another trading post, Fort Zealandia in Formosa, the present-day Taiwan. They had already wrested the cinnamon trade of Ceylon, now Sri Lanka, away from the religion-obsessed Portuguese. In 1641, the shogun of Japan allowed the VOC to set up an unfortified trading post at Deshima, a small artificial island in the deep-water harbour of Nagasaki, after the company had lent its cannon to help suppress a rebellion by Japanese converts to Catholicism in the nearby Shimabara Peninsula, reportedly shouting, "We're not Christians, we're Hollanders!" to explain their coldly commercial intervention. For the next two centuries, the Dutch – guarded like convicts, forbidden to bring their families, denied all but minimal contact with local people, mostly shopkeepers and prostitutes – comprised the only Western outpost in a Japan sealed off from the outside world. They were obliged, despite bitter protest, to trade at prices set by the shogun's men, their own tactic in the Spice Islands, to send a yearly delegation of merchants bearing tribute to the shogun's court at Edo, now Tokyo, and to perform comic dances to show how ridiculous foreigners really were.

Why did the Dutch endure these humiliating terms for a trade that never turned a profit? It was all part of the master plan of Jan Pieterszoon Coen. Inhospitable as they may have been, the Japanese were, as long as their mines lasted, the world's only source of silver not controlled by Roman Catholics and thus available to the Protestant Dutch, and silver was the one payment (until the British found opium) the Chinese would take for their tea, silks and fine ceramics, greatly in demand throughout Asia, particularly in Japan, with good future prospects in Europe as well. Governor-General Coen was planning to buy cheap in Asia and sell dear in Europe and the rest of the wealthy world by getting the Asian trade to finance itself. "I am of the opinion," he wrote to the seventeen High Mightinesses, the directors of the VOC, "that matters can be brought to a point that you will not be obliged to send any money at all from Patria", the latter the VOC's obsequious term for Holland, much as latter-day overseas executives might diplomatically email head office.

Never fully achieved, we can see here the ultimate dream of the Walmarts, Kmarts and the rest *de nos jours*. It was in pursuit of Coen's vision that his successor, Anthony van Diemen, in 1644 sent his most adventurous captain, Abel Janszoon Tasman, to see what deals the rumoured Great South Land might offer. Other VOC skippers, misjudging their longitude, had been wrecked or had close calls on the coast of what is now Western Australia where (little did they know) there was definitely no business to be done.

Tasman's sailing orders were to keep further south to the Roaring Forties and see where they took him. We know that this took him to the discovery of New Zealand before returning to home base in Batavia, missing Australia except for the fragment he called Van Diemen's Land. It was not until 1657 that the English East India Company, originally a coffee-house traders' clique, stole a leaf from the VOC's account book and began selling shares, not in single voyages but in the company itself. By then the VOC was easily the biggest and wealthiest commercial enterprise the fast-shrinking world had seen.

It was Dutch capitalism – if that term, yet to be invented, is appropriate. Daniel Bell, in his classic 1978 book *The Cultural Contradictions of Capitalism*, prefers "economism", arguing that the former Soviets practised it too, and we might call it "bottom-lineism". It certainly got off to a flying start, especially considering that the Netherlands' own products, gin, cheese and various cottage crafts – were not exactly in global demand. Amsterdam was

Europe's wealthiest trading city, Dutch wages the highest in the world. As mentioned before, Amsterdam's Beurs was the first continuously trading stock exchange; and in its first few decades, Dutch punters pioneered short-selling (ominously, in Dutch, *windhandel*), option trading, puts and calls, debt-equity swaps, merchant banking, unit trusts and the rest of the speculator's bag of tricks, much as we now enjoy them. With the first capitalism came its specialised offshoots: insurance, retirement funds and other attempts at risk reduction, the first asset-inflation bubble, the Tulip Mania of 1636–37, and even, in 1607, history's first bear raider, a sly ex-shareholder named Isaac le Maire who dumped his VOC stock, forcing the price down, and then bought it back at a discount. The breathless speed with which it all surfaced beside the Zuider Zee suggests that capitalism as we know it may be less the Devil's invention than what happens naturally when hard men who sincerely want to be rich meet to do business, free of all cultural, class or religious restraints.

To contemporaries, there seemed something magical about it and the Dutch of the Golden Age (1603–1689) were not the first people to mistake a temporary run of luck and favourable circumstances for the discovery of the Secret of Eternal Wealth and Happiness. The free-market fundamentalist's argument that market-based economies automatically generate liberal social institutions is more dubious. Rather, the Dutch experience suggests that a society newly freed of impediments, often as the result of external accidents like the defeat of the Armada, is then likely to generate both market economics and a liberal intellectual climate; but, as many examples from Franco's Spain to Pinochet's Chile suggest, markets can continue long after liberal ideas have been stifled. Holland's Golden Age certainly exemplified capitalism's first, fecund, stage.

The success of Jan Compagnie's freebooters in making Holland the world's corner grocery shop was soon confirmed on Europe's doorstep by the Dutch Republic's new navy, made possible by a certain Cornelis Cornelisz van Uitgeest's 1596 invention of the windmill-powered saw, easing the worst bottleneck to an expanding fleet, the sawing by hand of ships' timbers.

When King Charles lost his head in 1649, the new Lord Protector of the English Commonwealth, General Oliver Cromwell, tried to interest the Dutch in a common front of the two Protestant republics against Rome. The Dutch were as amused at the idea of an alliance with their main trade rival as the French were three centuries later when Winston Churchill offered them a political union with *l'ennemi héréditaire.* Instead, the English and Dutch fought three wars over trade and resources, provoked by English insistence that they owned the Channel and that the Dutch dip their ensigns to English warships.

As the Bible offers few useful tips on seamanship, Dutch maritime experience soon corrected the humiliation. In June 1667, the great Dutch Admiral Michiel Adriaanszoon de Ruyter sailed boldly into the Thames, burnt most of the English fleet, captured the flagship *Royal Charles* and towed her back to Holland, a broom lashed to his masthead signalling that he had swept the English from the seas. The diarist John Evelyn called it "a dreadful spectacle as any English man saw and a Dishonour never to be wiped off". When another Dutch squadron followed, Mr Secretary Pepys of the Admiralty, in his secret diary, quoted an English officer as wailing: "I think the Devil shits Dutchmen."

They were desperate days in England, coming on the heels of the Plague year, 1664 and the Great Fire of 1666, and the English were ready for any peace they could get. The Dutch terms were hard. In exchange for their trading post on the Hudson, Nieuw Amsterdam, where no groceries grew, the Dutch accepted Suriname, with slave-grown sugar, and Pulo Run in the Moluccas, sure they had the better of the bargain. Yet such are the vagaries of human affairs, within two decades the Golden Age was over and the VOC headed for ruin and a grim warning for our day.

What went wrong? The answer has long been hidden in scattered records in Dutch, a language not every scholar reads. In their pathbreaking study, *The First Modern Economy: Success, Failure and Perseverance of the Dutch Economy, 1500–1815*, (Cambridge University Press, 1997) Professors Jan de Vries of Berkeley and Ad van der Woude of Wageningen in the Netherlands have joined two potent techniques, the French Annales School of "eventless" daily life and American econometrics, to unearth a prophetic rise and decline. In 1689, the resolutely Protestant English Parliament, fending off a feeble attempt at a royal coup by the not-so-covert Catholic James II, invited the *stathouder*, William of Orange, and his wife (and cousin), Mary Stuart, James's estranged daughter, to ascend the English throne as joint sovereigns, William and Mary. Both were reassuringly rock-ribbed Protestants, but the invitation came with conditions: the pair were henceforth to have no power over the kingdom's finances, which would be voted annually by Parliament. Since then, English (soon to be British) bonds have been backed not by the good intentions of a municipality, the honest face of a merchant or the promises of a king, but by a body with power to raise taxes to redeem them. "Gilts" soon became the world's most sought-after investments, paying a sure six per cent, year in, year out. The long-term effect on the Dutch Republic's finances was fatal.

But, relates Simon Schama in his *The Embarrassment of Riches: An Interpretation of Dutch Culture in the Golden Age* (Vintage, 1987), ethical decay had already set in across the North Sea, instancing one of Rembrandt van Rijn's most celebrated works, *The Night Watch* (1642). From his mountain fastness in Geneva, Jean Calvin had taught that great wealth, a likely result of the idea that success in one's worldly calling was a sign of Divine favour, was a test of moral fibre. Between 1630 and 1680, the VOC cleared three million guilders a year in gold and silver from Asia to its shareholders in Patria. Dutch burghers, reports Schama, failed this "ordeal of prosperity", succumbing to the lure of pageantry and display, recorded by the school of painters who have made it immortal in their paintings on canvas with ground pigments spread in varnishing oil, a technique invented in the Netherlands with ingredients available in seaports.

In 1642, Rembrandt accepted a commission to paint a gigantic canvas, four and half metres by five, of such a ceremonial occasion, a public parade of the Militia Company of Captain Frans Banning Cocq, generally called *The Night Watch*. Captain Cocq's men, however, are not soldiers, but prosperous merchants from the drapers' quarter of Amsterdam, and look it – sleek, paunchy and hugely self-satisfied. As a painting it is magnificent, but the theme might well come from *Hello!* magazine or Life Styles of the Rich and Famous. Other paintings of the school show such upper-crust domestic tasks as weighing pearls and choosing silk gowns for chamber concerts. As Schama wryly observes, republics rarely live up to the innocence of their origins. If born in austerity, they invariably flourish amidst pomp. "At the zenith of its power and brilliance in the mid-seventeenth century, the Dutch Republic was no more immune than any other from acts of elaborate self-congratulation," he writes.

What the painters did not record, presumably because there was no market for them, were scenes out on the sea frontiers where the money came from. On VOC voyages to the Far East, two ships out of every hundred foundered on the way out, four on the return, and only a third of those who took to the sea ever came home. This left Dutch girls without husbands and exhausted the supply of adventurous young men from the smaller Dutch towns. Immigrants flocked in, mostly from Germany, depressing Dutch wages as the Dutch birthrate began to falter. Competitors, notably the British, eroded VOC profits, while the wealthy began to invest in British gilts or in Amsterdam real estate, which they could look at every day. The bold investors of the republic's early days had become risk-averse rentiers obsessed with showing off their wealth (Dutch historians call the later republic The Periwig Age). Once-united Netherlanders divided into classes, the have-much and have-little, would-be reformers and last-ditch defenders of the status quo.

The VOC's finances faltered but Jan Compagnie was so central to the economy that it was nationalised and blundered on as before. As significantly, Holland stalled on the technological frontier that had opened up across the North Sea, with James Watt's 1769 patent of the separate-condensing steam engine. Dutch drainage boards were among the first buyers of steam engines, and Newcastle coal cost the same in Rotterdam as in London. But the wind saws, the peat fires and their owners were too well established, and the Dutch system of loosely co-operating vested interests, which had been its strength, was unable to meet the new challenge. Schama estimates that before the steam engine, the Dutch were already using more energy per head than anyone else, but it was only a way station. Simultaneously assailed by reformist Patriots and invading French revolutionaries, the Dutch Republic, unmourned and awash in public debt, collapsed in 1795. It was "rich without being prosperous", an economy going nowhere. With Napoleon Elba-bound, in 1814, the despondent Dutch recalled a prince of the House of Orange from exile in England to be their first king, inevitably Willem I. The next half-century was rough, as the modern Dutch nation struggled out from the wreck of a glorious, and for us a most instructive past.

De Vries and van der Woude conclude about modern economic growth, on the Dutch example: "It is not self-sustained, exponential and unbounded and – not to mince words – the [Dutch] Republic's pioneering experience in the sixteenth through eighteenth centuries, including its experience with stagnation, may end up being a fair model for the process begun in most Western countries some time between 1780 and 1850."

Now let's see. Two centuries on from 1800 brings us to about now. This is getting rather close to home – a warning, if not from Abel Janszoon Tasman himself, then from two of his most scholarly, insightful compatriots.

It would be unfair, or fairly unfair, to present all the Dutch of the Golden Age as snobbish, conscienceless money-grubbers. Dutch humanism has left us a legacy equally important to the state we're in now. Desiderius Erasmus of Rotterdam (1469–1536), the illegitimate son of a Catholic priest and a physician's daughter, later a priest himself, was a wandering scholar who visited Renaissance Italy, France and Tudor England, listening with a good-humoured "Dutch Ear" (his term) to the theological disputes of the oncoming Reformation. Erasmus rejected both Martin Luther's doctrine of predestination and Rome's claim to total authority, arguing that no human can claim infallibility, urging the disputants to use commonsense and, as we should say, to lighten up: "Even the wise man must play the fool if he wishes to beget children," he wrote in *In Praise of Folly*. The Dutch seaports were havens of toleration and diverse thought. Protestant Huguenots expelled from France flourished, as did "New Christians" (actually Sephardic Jews) from Spain and Portugal, and later Ashkenazi Jews from Germany. The painter Jan Vermeer lived and died a Catholic in Delft. The cosmopolitan

Dutch ports even tolerated a measure of intolerance, a difficult test: the sublime philosopher Baruch de Spinoza (1632–77), who combined a deeply spiritual outlook with the questioning of all dogma, was excommunicated by his Amsterdam synagogue for (Jewish) heresy and became the first person in Europe without a religion, and lawyer Hugo Grotius (Huig de Groot to his neighbours in Delft, 1583–1645) laid the legal basis of freedom of the seas, peaceful trade between nations and just terms for the defeated in war, much as we apply them in theory now.

There was a living link between the Dutch Golden Age and our own outlook, and we know his name. Bernard de Mandeville was born in Rotterdam in 1670, studied medicine but was drawn to a big-city career as poet, wit and controversialist. Wandering the *bordeel*, Amsterdam's red-light district, he had a shrewd insight: the whores exhibiting themselves through the windows of their houses in Vermeer-like interiors were the mirror image of respectable Dutch domesticity and, in fact, made it possible for virtuous women to live in safety a few streets away from the docks where sex-starved sailors landed after long, dangerous voyages from the Far East. His essay, "A Modest Defence of Publick Stews", later got him into trouble with the prim London magistrates, but he had undoubtedly described a sad, sordid aspect, then and now, of every busy port city of the world.

In 1699, de Mandeville moved to London to learn English, married an Englishwoman and fathered two children. He also developed a taste for English satire, the tradition continued today by *Private Eye*. This inspired 434 lines of engaging doggerel called *The Fable of the Bees*, or, *Private Vices, Publick Benefits*, which was denounced as undermining public morals and, as a result, went through many editions with Mandeville's replies to his shocked critics, who nicknamed him Man-devil.

It tells of a wealthy and powerful beehive whose inhabitants act only in pursuit of gain and fame. Nevertheless, they espouse an ethic that condemns this behaviour and frequently lament that their society is full of sin. Irritated by their constant whingeing, their god decides to make them all virtuous. In a flash, their prosperity collapses: commerce and industry fade away, and the bees leave their once flourishing hive and withdraw to live austerely in the hollow of a tree. The moral is that virtue can only lead to a poor, ascetic society, whereas the vices are the necessary engines of a wealthy and powerful beehive/nation.

The entomology of this is dubious – real beehives are more like absolute monarchies, if not prototypes of globalised production – but de Mandeville makes some shrewd points, overstated for satiric effect. Hypocrisy and self-righteousness do often go together, public philanthropists are trying to buy good reputations, few of us are as high-minded as we like to believe, and so on. De Mandeville was trying to shock, and nowhere did he succeed better than among Scottish intellectuals, haunted like the Dutch by a stern Calvinist inheritance. Frances Hutcheson, professor of moral philosophy at the University of Glasgow, wrote a fiery book denouncing de Mandeville.

But Adam Smith, Hutcheson's star student and Glasgow successor, took what at first sight seems a different, even a Mande-villainous view. In his book *The Theory of the Moral Sentiments*, Smith followed his teacher in holding that "virtue is upon all ordinary occasions … real wisdom and the surest and readiest means of obtaining both safety and advantage". But, Smith wondered, how could a society function when people were no better than they should be, or even not as good? To find out, he undertook fieldwork, or rather tavern work, among Glasgow tobacco merchants – hard, unsentimental men. The result was the most famous of economics texts, *An Inquiry into the Nature and Cause of the Wealth of Nations*, which introduced the metaphorical, if not mystical, notion of "the invisible hand", mentioned in only one passage: "Every individual necessarily labours to render the annual revenue of the society as great as he can. He generally, indeed, neither intends to promote the public interest, nor knows how much he is promoting it. By preferring the support of domestic to that of foreign industry, he intends only his own security; and by directing that industry in such a manner as its produce may be of the greatest value, he intends only his own gain, and he is in this, as in many other cases, led by *an invisible hand* to promote an end which was no part of his intention. Nor is it always the worse for the society that it was no part of it. By pursuing his own interest he frequently promotes that of the society more effectually than when he really intends to promote it. I have never known much good done by those who affected to trade for the public good."

Or, more succinctly: "It is not from the benevolence of the butcher, the brewer or the baker that we expect our dinner, but from their regard to their own interest. We address ourselves not to their humanity but to their self-love."

Those who have not waded through *The Wealth of Nations* are often convinced that it rewrites Mandeville in a Scots accent, but this is far from true. Wondering why Scotland, united with England since 1707, was poor – at least in comparison with its southern partner – Smith hit on the division of labour that was made possible by England's far bigger internal market. The biggest manufactory ("factory" for short) he cites is an enterprise making pins, which employs nine workers, but his reasoning is sound: neither they nor their employers can live on pins, but must exchange them for the products of farms, forests or other factories. The secret of wealth therefore lies in trade, not in governments amassing bullion, which Smith denounces as "mercantilism". As he knew, all known human societies in touch with their neighbours engage in trade, and many have hit on money in some form or another to avoid having to lug their products to market and to act as a store of value between times. Trade does not work on elevated moral principles, but on the meeting of willing sellers and willing buyers.

The "invisible hand" was, in fact, no more than an aspect of human behaviour, long known and not at all mystical. Smith himself lived by lofty moral principles, never traded a plug of tobacco in his life and had a low opinion of those who did, and of employers generally: "People of the same trade seldom meet together, even for merriment and diversion, but the conversation ends in a conspiracy against the public, or in some contrivance to raise prices ... We rarely hear, it has been said of the combination of masters, though frequently those of workmen [aka trade unions] but whoever imagines upon that account that masters rarely combine, is as ignorant of the world as of the subject." Smith's remedy for these evils was not the invisible hand of self-interest but vigilance by the authorities to break up monopolies, an independent judiciary, free public education for poor adults and a strong military.

Nevertheless, long before *The Wealth of Nations* was published in 1776, his theory had been exploded in its only practical test, the Dutch Republic. De Mandeville left Amsterdam in 1699, when the first modern economy was in its Golden Age and everything seemed possible. Adam Smith pursued his philosophical studies in France but never visited, much less studied, the republic as it tottered towards collapse. The Dutch merchants had pursued their self-interest, as they saw it, resulting in a divided society, mountains of debt, a real estate bubble and an obsession with ostentatious luxury, particularly in displays of exotic food and drink – times, come to think of it, disturbingly like our own. Who could have suspected, as Abel Tasman set off on his long business trip, that the world would one day be conquered by an economic system invented by a crowd of merchants haggling on an Amsterdam fish wharf, in the rain?

Did the Dutch-invented system we call "capitalism" create our modern fossil-fuelled world economy, or did fossil fuels make the Dutch system viable? The historical record is clear: they grew together, one reinforcing the other. In 1776, coincidentally the year *The Wealth of Nations* was published, the future firm of Boulton and Watt, machinery manufacturers of Birmingham, England, installed its first two commercial steam engines, one to pump out a coalmine in Staffordshire, the other to power the blast furnaces of the pioneer ironmaster John Wilkinson, whose name is remembered by a brand of razor blades. Watt had actually built the first working model of his separate condenser engine in 1765 at Glasgow University, where he was a friend of Adam Smith, but it remained a scientific toy until Watt met Matthew

Boulton, owner of the Soho Ironworks in Birmingham, who saw that Watt had found a relatively efficient way of capturing the energy released by burning coal for productive purposes, among such purposes being the mining of more coal. Watt's was not the first device to turn heat energy into motion but it was by far the best up to his time and, as important, the first to be made and sold on capitalist lines. Boulton borrowed £10,000 from the London bankers to perfect and market Watt's idea, and both sides made a tidy profit. Watt and Boulton became wealthy men, loaded with honours, and towards the end of his life the Scot Watt even began to take holidays.

Generations of students have learnt the sequel, the triumph of progress and (on the whole) peace of the nineteenth, the British century – the happiest time, it was argued, that humanity had up to then known. No one was more dazzled by capitalism's achievements than the fiery young scholar Karl Marx, writing his *Communist Manifesto* in London in the year of revolutions, 1848: "The bourgeoisie [by this fancy French term Marx meant that not the urban middle class, but anyone with money to invest, like his wealthy patron and collaborator Friedrich Engels] during its rule of scarce one hundred years has created more massive and more colossal productive forces than have all preceding generations together. Subjection of nature's forces to man, machinery, application of chemistry to industry and agriculture, steam navigation, railways, electric telegraphs, clearing of whole continents for cultivation, canalisation of rivers, whole populations conjured out of the ground – what earlier century had even a presentiment that such productive forces slumbered in the lap of social labour?"

The manifesto goes on to express, in billowy language, Marx's sharpest insight: there can be no rest for a system that depends on never-ending change and expansion: "The bourgeoisie cannot exist without constantly revolutionising the instruments of production, and thereby the relations of production, and with them the whole relations of society ... All fixed, fast, frozen relations, with their train of ancient and venerable prejudices and opinions are swept away, all new-formed ones become antiquated before they can ossify. All that is solid melts into air, all that is holy is profaned, and man is at last compelled to face with his sober senses his real conditions of life and his relations with his kind."

What, we may well ask, is the attraction of this non-stop, wearying technological razzmatazz? The best answer comes, predictably, from a romantic poet, Lord George Byron, who describes to a friend what still sounds like a night at the trots, a Sunday paper or an evening's TV.

The great object of life is Sensation – to feel that we exist – even though in pain – it is this "craving" void which drives us to Gaming – to Battle – to Travel – to intemperate but keenly felt pursuits of every description whose principal attraction is the agitation inseparable from their accomplishment.

We live, in more with-it language, for kicks, and supplying them is yet another ever-expanding business.

There is no simple, cause-and-effect relationship between the Industrial, or hydrocarbon, Revolution and the explosive growth of human population. As with its physical counterpart, the climate, the neat notion of cause and effect is not an adequate tool for analysing phenomena with complex inputs; causation is a human invention, made in the Stone Age, to enable us to get a handle on the way the world works and plan tomorrow's activities, and it still serves us well enough in daily life. In strict truth, things are interconnected and everything causes everything else, but to a greater or lesser extent.

Scientists have long been accustomed to expressing their conclusions in terms of the probability that a given set of conditions, all other things being equal (which they never are), will have an estimated range of consequences, so that the scientists of Working Group One of the Intergovernmental Panel on Climate Change reported from Nairobi in 2001 a sixty to ninety per cent chance that the CO_2 concentration in the air had not been higher in the past twenty million years, which is yesterday in geologic time, and the equally star-studded scientists of Working Group Two offered similar odds that by the year 2100, which some of us may well see, the global average air temperature will be up by between 1.4 and 5.8 degrees, from rather warm to a real scorcher, and mean sea level up from 0.09 to 0.88 metres, the difference between a puddle and a pond. The scientists are only trying to err on the side of caution, but to the rest of us this can sound like the sort of sporty wager Lord Byron recommended to keep life interesting, and maybe it will, the way we're going.

That said, we have begun to see a pattern of sorts in the stages by which the industrial system, aka hydrocarbon civilisation, is still taking over the world. It bears a family resemblance to the British prototype, which, being the first, cannot have been imitated from anywhere, except in its financial arrangements from Holland. The sequence goes thus. First, improvement in agriculture creates a food surplus. This can follow the introduction of new crops, but the most favourable condition is peace, in which "best practice" spreads among farmers, most of whom have long been accustomed to growing food for a parasitic military class.

Next, or at about the same time, a distribution network is created or refurbished. Water transport is still the most energy-efficient, which is one reason the original Industrial Revolution began in the British archipelago, and in

Asia in the Japanese. Britain acquired inland waterways in the early 1700s, climbing over hills by means of pound locks, long in use in China, and financed by the Dutch system of selling shares. One strong horse could pull an eighty-ton load. Transport systems need labour to build and enlarge markets; the Grand Union Canal, for instance, made a port out of centrally located Birmingham.

Next, a source of fossil fuel is required. Marco Polo, visiting Cathay (North China) in the 1280s, noticed that the Chinese heated their homes by burning black stones, but no one in Venice believed him or saw the point. Captain James Cook learnt his seamanship in the 1750s as mate of a collier hauling coal from Newcastle, England, to Bristol, and, remembering the blue line of mountains and low shore of South Wales as seen from across the Severn Estuary, named his discovery in the South Seas twenty years later New South Wales. Cook's coal had been destined for the many small ironworks that had sprung up in the hilly Severn area using the local streams for power, and the operator of one of them, Abram Darby, had already succeeded in smelting iron with coke, in place of the charcoal which had used up most of England's remaining forests. The Darby family were "weighty" (strict) Quakers kept out of the profitable casting of cannon by their pacifist views. Darby visited Holland and brought back Dutch workers to demonstrate their iron-casting skills, which he adopted for making pots and pans. The first stage of the first industrial revolution was completed when George Stephenson's Rocket hauled the world's first steam-powered passenger train from Stockton to Darlington, near Durham, in 1825. Thirty years later, in 1855, Australia's first steam-powered railway ran from Sydney to Parramatta. Hydrocarbon civilisation had transported itself halfway around the world.

The population surge of the nineteenth century owed something to all these developments. The new railways made it possible for farmers' sons and daughters to flock to the new industrial cities. But, for a generation or two, they followed the reproductive pattern they had learnt on the farm, where children were hands to work and, if they survived, insurance against old age. Living conditions in the cities were harsh, but nevertheless they expanded, and the new moneyed middle class took to sending their sons not to the Army or the Church, but to private schools and universities where some took up science or medicine, with great benefit to public health. The railways ended the famines that had kept populations in check, except in Ireland, which had become dependent on a single crop vulnerable to disease, the potato. Famine emigrants swelled the labour forces of England, Wales and Scotland. Britain became the world's leading exporter, not only of power-loomed textiles but of technology, coal, capital and people. By 1900, one Englishman or Welshman in ten was a coalminer, as the world was drifting, suspected by few, into a titanic, industrialised war over access to resources. Marx, in his role of failed prophet, considered that such "imperialist" wars were impossible, but when the guns began to fire, the new industrial proletariat of Europe, grown relatively prosperous since 1848, patriotically flocked to join up. What had made hydrocarbon civilisation and thus industrialised war both possible, and inevitable?

No human eyes saw the formation of the fossil fuels that have made our hydrocarbon civilisation possible and its future problematical. It was the one-off event, never to be repeated, that geologists call the Carboniferous Age. It began 360 million years ago and ended seventy-four million years later. In that (relatively) short time, all the coal we use so profligately and the oil we pursue so voraciously was laid down, once and for all time. Fourteen million years' worth of that undeserved legacy went up in smoke in the last century alone. Our own fair-weather species was far in the future when it was laid down but we can see the circumstances that produced our coal, oil and natural gas, and begin to understand why it will never happen again, all from a deckchair on a comfortable river cruiser a few hours away from Strahan, a journey that will take us past Sarah Island, the short-lived convict settlement in Macquarie Harbour, along the Gordon River and deep into the Tasmanian wilderness, the last stand of the vegetation of our lost southern super continent, Gondwanaland.

Trees and plants have primitive ancestors, just as we do, and the green walls we see on either bank are living fossils, Gondwanaland's last stand against what is our still-evolving world. If you look closely at the celery-top pines gliding by, you will see that they have no leaves but rather the feathery fronds from which leaves descended aeons ago. The flowering swamp gums, fifteen metres high, are not true trees but the world's tallest plants. The river under you is stained dark brown by button grass, topped with foam from the rapids higher upstream, and looks exactly like Guinness stout, but is pure to drink. Then, as you leave saltwater behind, you see Huon pines, surely the strangest of all living trees. To defend itself against insect and bacterial attack, the Huon pine relies on oil, which makes it practically worm- and rot-proof, ideal for boat-building – the reason convicts were sent to bleak Sarah Island to cut it. Slow-growing – some are 3,000 years old – Huon pine has closed-grained yellow wood, showing fine detail in carvings. Felling is forbidden, but convict-cut trunks and branches that have lain in ooze since the 1830s are still found, perfectly preserved, and turned into handy bowls and cheese boards.

The Huon pine is not endangered – there are seedlings by the thousands scattered through the wilderness, which will mature four or five hundred years hence. At Heritage Landing, well up the Gordon, tour boats moor at a boarded walk into the rainforest, with 2,000-year-old Huon pines close enough to touch and all but impassable ferns, mosses and creepers below our feet. A seaplane from Strahan's wharf lands higher up the Gordon, near its junction with the Franklin, and also accesses a boarded walk on the wild side of the rainforest. Our world, the only one we have, was a damp, wholesome place millions of years ago, as we can see. Drink in hand, diesel throbbing below, I'm not sure we've improved it much.

The Carboniferous Age can never be repeated. Evolution (or some, following the Prophet Job, might say the Lord) has given and evolution/ the Lord has taken away. In 1916, the prophet of the theory of continental drift, the German geologist Alfred Wegener published what then seemed a far-fetched explanation of a fact first noticed by medieval map-makers, namely that the bulge of eastern South America tucks neatly under the armpit of West Africa. Wegener surmised that they had once been joined, and that all the other landmasses had been linked as well in a huge single continent he called Pangea (Greek for "all lands"). This had first split into two, a southern super continent which Wegener called Gondwanaland, after a place in India, composed of the present Australia, New Zealand, the Arabian Peninsula, Africa, India, Ceylon and South America, all centred on Antarctica, and a smaller northern landmass he called Laurasia, which included the present Europe, Greenland, Siberia, North America, Kazakhstan and North China. Since then, he theorised, the continents as we now fly over them had been drifting apart and occasionally running into each other, like lost surfboards. But what were they drifting on? Not the oceans, which simply gushed in to fill the gaps. The only other possibility was the core of the earth itself, known from volcanic eruptions to be liquid and very, very hot.

The world had other things on its collective mind in 1916 and Wegener's idea attracted little support. A true hero of science, he died in 1930 at fifty, on an expedition to the North Pole to look for more evidence. It was not until the 1960s that confirmation began to flood in from a dozen different sciences that Wegener had been right. Under the name "plate tectonics", the theory has been immensely fruitful and is all but universally accepted. And, although we cannot see coal and oil forming, much less make any more, the process is now well understood.

Life as we know it is not made possible by supermarkets, but by the green pigment called chlorophyll, which uses the energy of sunlight to synthesise carbohydrates, organic compounds of carbon, hydrogen and oxygen.

The last two are the components of water, so carbohydrate means "watered carbon". This process began under the sea, protected from the sun's most destructive radiation by a layer of water and, as a souvenir, our blood has almost the saltiness of our marine mother, and the haemoglobin that carries oxygen to our muscles is a close chemical relation of chlorophyll.

Seaweeds and algae are still hard at it in the upper layers of the sea, but when dry, or dryish, land eventually surfaced above the waves, green-tinted vegetation soon followed, if a billion years is "soon". Free oxygen is a product of splitting the hydrogen and oxygen of water, so in a geological trice we had both green vegetation to add oxygen to the air and creatures to breathe it, starting off with our little friends the insects (little because their system of absorbing oxygen through many small external tubes severely limits their size).

At this point, the world entered the Carboniferous Age. In some ways it was curiously like our own. Average temperatures and humidity were much like ours, both considerably lower than before or since. A landmass (Pangea's successors) stretched from pole to pole, blocking the natural circulation of wind and currents from west to east (because the planet turns the other way) and producing such odd eddies and oscillations as the El Niños and Gulf Streams of those times. $C0_2$ in the atmosphere averaged lower than ours, around 300 parts per million, while oxygen was higher, between thirty and forty per cent, making possible strains of giant insects looking for vegetation to supply their daily carbohydrates. The big difference was in the vegetation, the pre-trees we have already glimpsed in the Tasmanian wilderness. The interaction of insects, fungi, bacteria, climate and continental drift produced coal, oil and natural gas, a combination of circumstances now well understood.

These fossilised fuels are essentially carbohydrates that have lost their oxygen, making them more concentrated sources of energy when dug up and recombined with the oxygen we still have left in our air.

The process of capturing and preserving ancient sunlight for our future use was, at best, chancy. The simple vegetation of Gondwanaland and Laurasia grew in swamps and along the courses of slow, meandering rivers.

When sea levels rose or lake levels subsided, the vegetation was drowned, preventing bacterial action. Titanic floods or marine inundations covered it with layers of silt, sometimes thousands of metres deep, where heat and pressure compressed the swamps into coal measures, sometimes inches, sometimes tens of metres thick. The timing had to be just right – if the water rose too fast, no worthwhile accumulation happened; too slowly, and the swamp was uncovered and dispersed.

Australia was well placed, having in those days a very wet interior and being close to the centre of the action, the always-frozen Antarctica, which is why Chinese coalcarriers now queue outside Newcastle and Gladstone. The formation of oil was even chancier, which is why there is much less of it to fight over. Coal stays where it is, under overlying sandstone, limestone or shale, depending where "is" is, until someone digs it up or rips off the covering layers. Oil started off like coal in drowned vegetation but, before it could solidify, migrated through nearby rocks that happened to be porous, to a reservoir, usually fifty metres or so thick (an "oil window") solidly sealed under an impervious cover that can be kilometres deep. For reasons not clearly understood (except by those who see the hand of Allah), but possibly connected with the ice ages that continued through the Carboniferous Age, the flat land and immense rivers of the future Arabia saw a lot of oil formation, some three-quarters of the world's known reserves. Natural gas (methane or swamp gas) followed a similar course, and is held in equally deep, impervious and rare reservoirs.

Why did it end? The explanation can be glimpsed in the Tasmanian wilderness, one excellent reason for its being declared a World Heritage Site in 1982. Lacking a true sap system, the ferns, mosses, plants and creepers of the wilderness, or temperate jungle, could never stray far from running water, which demanded constant, heavy rainfall (up to two metres a year in many places). It was not until true trees, with bark and leaves, evolved that terrestrial vegetation was able to survive drier, colder conditions. Bark is mostly lignin, a tough substance that resists fungi, bacteria and insects. The resulting forests are hardier and more dispersed and, if inundated, simply return their carbon to decay underground, unavailable for future use.

Why, then, does the Tasmanian remnant of Gondwanaland survive, resisting the invasion of all but a few "modern" trees? For the same reason that locates Cape Grim, and Tasman's landfall, further up the coast. All are well within the Roaring Forties, the winds that circled the planet as they blew around Pangea, when it was our sole continent. For proof, there are some inaccessible islands off southern Chile. They were once our neighbours in Gondwanaland, and they too preserve something of its ancient vegetation.

But evolution is a one-way street: there is no way back.

The same applies to us, anatomically modern men and women, *Homo sapiens sapiens* ("the wisest of the wise") as we call ourselves, *Homo smartypants* as we might be better named, being too clever for our own good.

Charles Darwin's guess that we originated in Africa has since been supported by such an avalanche of fossil finds as to become accepted wisdom, although, being the argumentative creatures we are, some will always be found to disagree, which is as it should be. There is little serious argument, however, about naming the first of our line to trudge out of Africa. He (embracing she) was *Homo erectus*, upright man, and hit the road some two million years ago. While not, perhaps, a type we would invite home to dinner, *Homo erectus* is an ancestor we can take pride in. He stood about 180 centimetres tall, his lady a few centimetres shorter, and both had massive eyebrow ridges and heavy jaws (they ate their food raw, at least in their early days). As the poet/philosopher Ogden Nash put it:

They had an aspect somewhat simian;

And not like regular men and wimmian.

Pin-ups or not, *Homo erectus* still had formidable achievements. They survived at least a dozen glaciations and hot interglacial spells, where we are getting windy about one. They made simple stone tools. They pioneered the human uses of fire. They had the vocal equipment for speech, on the high-pitched side, more Boy George than Peter Dawson. They spread throughout the Old World, the first of our kind to do so, although they seem to have done no boating and never made it to Australia or the Americas. They did, however, leave their remains in Indonesia (Java Man), North China (Peking Man), Africa (Turkana Boy) and other places, carbon dated at up to 1.8 million years old. As the aeons rolled by, the original *Homo erectus* evolved into sub-species, Neanderthal Man and Heidelberg Man being the best-known among Europeans. What kept them on the move? The curiosity we still have about what lies over the next hill. And what happened to them? We're not sure.

What we do know is that our own kind, *Homo sapiens sapiens*, took the well-trodden trail out of Africa a bare 120,000 years ago and, in that blink of geological time's eye, has taken over the world and is looking for new ones to conquer. What did/do we have going for us? Certainly not physique. Our brains, however, are twenty-five per cent bigger than stolid old *Homo erectus*'s, pushing our physiology to our design limits. The human birth canal cannot get much wider without compromising a mother's walking ability to keep up with the group, meaning that we are born even more helpless and our full growth has to be postponed to the years of adolescence. Our jaws are so weak that we cannot get food down without cooking or grinding it. What's in our big brains? Not simply speech – *Homo erectus* had that – and our four-footed mates from the Stone Age, our dogs, can communicate their feelings to us quite well – but abstract speech, the ability to manipulate reality in its absence, to plot, plan, tell stories, reflect on yesterday and tomorrow and eventually to get to the Moon.

We know that we shared the Earth for something like 60,000 years with more sophisticated later versions of *Homo erectus*, although the periods of close contact seem to have lasted for a thousand years or less before, as far as we know, the latecomers took over. What happened to the losers?

It used to be thought that we interbred with them, more along lines of kidnap and rape than moonlight and roses, and that this violent intermixture might account for the geographical concentration of humans of different outward appearance, commonly – if erroneously – called races. An old-school anthropologist, the unfortunately named Carleton S. Coon, in 1962 published a book *The Origin of Races* in which he identified five: Australian, African Bushmen, Caucasian (a term which survives as a police euphemism for white), East Asian and African. Terms like Middle Eastern, Mediterranean or European had no part in his system and Coon conceded that, as far as we know, all living humans are mutually fertile, implying close matching of genetic inheritance. Coon did his extensive fieldwork in the 1920s and 1930s, but in the shadow of the Second World War, the Holocaust and Hiroshima, many scientists responded with a darker view of the human personality. Based on a proposal by the Australian anatomist Raymond Dart, the idea of "man the killer ape" found wide credence. A 1961 book by Robert Ardrey, *African Genesis*, ringingly begins, "Not in innocence, and not in Asia was mankind born." As to our *Homo erectus* rivals, maybe we just did them in. This would certainly account for our racism, in the sense of hostility and contempt for people of different appearance, and for the synthetic racism of beards, badges and uniforms to distinguish otherwise identical-looking friends and foes.

This controversy was raging (it rages still) when a new sensation came out of Africa. In 1987, an international team headed by the New Zealander Allan Wilson, following the then-new discipline of molecular biology, published an intriguing article in the scientific journal *Nature* theorising that all living humans were descended through a maternal line from a common female ancestor, who was inevitably nicknamed Eve. In Japan, where the theory became the basis of a hit TV series, she was renamed Masako, meaning "feminine elegance". The theory does not imply that Eve was a single buxom proto-beauty, rather that *Homo sapiens sapiens* started off with as few as 10,000 breeding pairs in the cool highlands of East Africa some 150,000 years ago.

From then, our history, written in the gases trapped in the ice cores from Greenland and Antarctica, shows that no new species debuting on the world stage has ever had it so good. About the time the boys were courting Eve, no doubt with more slap than tickle, we went through our first and only hot spell, quickly followed by a shortish interglacial, another slow cool-off and then the longest interglacial, some 60,000 years, that the record has to show, so benign from our point of view that scientists call it the "interglacial holiday".

Then, about 20,000 years ago, we had the one and only glacial period we have had to weather, with low CO_2 concentrations, high ice and low seas. It was, significantly, the icesheets spreading over Europe that finished off the last of our late *erectus* rivals, *Homo neanderthalis* and his rough-hewn cousins.

There is much evidence that agriculture began in the Middle East, somewhere north of present-day Iraq, on the fertile margins of the retreating icesheets. Yet the trove of Middle Eastern myths and legends that survive in Hebrew, and thus also Christian and Muslim scriptures, while they relate many floods and some droughts, never mention sheets of ice. An explanation leaps from the records: while *Homo erectus* battled the ice, we were living in the balmy climate, with no need for the insulation of clothing, remembered as the Garden of Eden. No wonder our myth makers thanked our creator, the divine prototype of all of us *Homo sapiens sapiens*.

Naturally, being human, we ascribe our good luck to good management. A better explanation was provided by the Serbian astronomer Milutin Milankovitch, who identified three irregularities in the Earth's yearly trip around our private star, the Sun, and daily revolution on its own axis. Briefly they are: Earth's orbit, which is not circular, but elliptical; its tilt from the vertical; and its axial precession, known in the trade as the "Milankovitch wobble". It so happened that all three have, in our time, been near the centre of their ranges. But the cycles go on. Thanks to axial precession, the seasons wander the globe and Sydney should have a winter Christmas some 12,000 years hence, assuming that either is still remembered.

Other elements that seem to have been working for us have been a relatively quiet time in gaseous eruptions on the Sun's surface, no known major strikes by asteroids, the absence of earth-darkening volcanic blow-offs and a favourable placing of our sun and its satellites in our own galaxy.

Closer to home, Milankovitch offered an explanation for the glaciation ("ice age")/interglacial cycle that in the past couple of million years seems to have settled into a ragged cycle of around 100,000 years. On this count, we are due, if not overdue, for a hot time around now; the Earth should be warming, as his theory predicts, the questions being: how fast? and by how much?

To be sure, we have not helped matters by ripping holes in the atmospheric security blanket over our heads, nor by using our fossil fuel windfall to go forth, multiply and get rich, to the point that by 2040, on some respectable estimates, we will need a second planet Earth to support us all on a sustainable basis. As our daily serving of climate-caused disasters reminds us, we are being forced by population pressure and the need to fill our stomachs to live in places a more prudent species would take care to avoid, magnifying the cost of more or less routine bad news and the speed with which we now see it.

Lifeboat Earth is dangerously (criminally if we were paying passengers) overcrowded as we navigate the stars, and more than a little sky-sick. What, on past planetary form, should we expect to run into, some foggy night?

The track record of our planet, forever set in stone and sand, warns us unequivocally to expect the worst. Of all the species that have ever lived, 99.9 per cent are extinct, including all our close relatives and such cutting-edge specimens (for their day) as the sabre-toothed tiger and the well-padded polar dinosaurs who used to work the area yet to become fashionable beaches near Melbourne. The past 450 million years have seen five mass extinctions, ascribed to various causes that have not gone away: bursts of cosmic rays, super-volcanoes, colliding continents. In the most severe, the Permian–Triassic Event, ninety-five per cent of all marine species and seventy per cent of land species perished, clearing the way for relative newcomers, the dinosaurs, to become kings of Castle Earth. They, in turn, got their come-uppance in the Cretaceous–Tertiary Event, sixty-five million years ago, when fifty per cent of species went west with them. The K–T Event, in palaeobiologists' shorthand, is widely believed to have been the consequence of the impact of an asteroid in the vicinity of the Yucatan Peninsula in Mexico, leaving a crater some thousand kilometres wide, which can still be traced.

These mass extinctions were in times of darkened skies, boiling heat, glacial cold and other climatic conditions sure to be fatal to a fair-weather species like us, so picky about our environment. The conclusion is clear: one of these days we, too, are due for the celestial chop, the only boxes to be checked being when and in what catastrophe.

In fact, many scientists argue that the Sixth, or Holocene (the geological period in which you are reading this) Mass Extinction is already well under way. In his famous 1975 book, *Sociobiology the New Synthesis* (Harvard University Press, reissued 2000), the Harvard biologist Edward O. Wilson calculates that we are now losing 30,000 species a year, which your calculator will tell you is three gone every hour. Wilson lists the reasons: landscape change, the over-exploitation of species, air and water pollution and the introduction of new species. These are due he says to people (us) who "excuse their gluttonous behaviour in crushing the planetary life support system".

All of these are the work of *Homo sapiens sapiens* and began not with the Industrial Revolution or with the invention of capitalism, but with the introduction of agriculture some 11,000 years ago, dividing the natural order into "crops"(useful to us) and "weeds" (to be eliminated).

Like the doomed Dutch Republic before us, however, we have gone much too far down our primrose path to even think of turning back. Nor, in our hearts or heads do we really want to. Just as the Golden Age of the Dutch left us a priceless legacy of humanism, reason and toleration as well as the businesslike explorations of Captain Tasman, so fossil fuels have brought us the sobering knowledge that we are mortal, as a species as well as individually, new forms of art and literature, longer, more comfortable and secure lives for more of us, and (perhaps not such a notable achievement, this) the up-ended packing-case CBDs of our great trading cities. To these achievements we can perhaps add political democracy. When humans are useful only as muscle-power, no one really cares about what is in our heads. Only when we become consumers are we likely to be consulted as voters. By coincidence, the slave-owner Thomas Jefferson's *Declaration of Independence,* which at least uses the language of democracy, was issued in 1776, the year of *The Wealth of Nations* and first sales of the Watt steam engine.

But "capitalism" and "democracy" are not identical twins: one is an economic order, the other a political system. Above all we now have the growing realisation that, sink or swim, we are all in this together, to the bitter end. So how to proceed?

One approach that has run into the sands is the 2001 Kyoto Protocol, already destined to join such high-minded curiosities of history as the Kellogg–Briand Pact of 1928 that outlawed war forever, and was in hindsight more of a hope that the winners could hold on to their gains without having to fight for them. The joint work of hard-fisted deal-makers and unworldly visionaries, Kyoto began with the idea that the market mechanism that is pushing fossil-fuel consumption, air pollution and, indirectly, populations ever upward could, with suitable tinkering, be redirected to slowing and, with luck, even reversing these ominous trends.

Kyoto, like Kellogg-Briand, fell at the first hurdle. It divided the world, whose common problems it supposedly addressed, into two: the rich nations ("Annex 1" in tactful Kyoto-speak) and the rest, the developing nations, eager to burn fossil fuels as fast as they can, and exempt from Kyoto restrictions. As the atmosphere doesn't care who pollutes it, this was inherently absurd, compounded by allowing the "Annex 1" nations of Europe to erect their own mini-Kyoto, the so-called "bubble" that allowed poorer European Union nations, such as Ireland, Spain and Portugal, to exceed their carbon quotas as long as the "bubble's" books balanced which, it is already clear, they will not. Kyoto's fatal defect was indicative of much to come: the climate problem is the world's but the effective players are nation-states, with completely different imperatives: polls, elections, national rivalries and, when all else fails, our large, fossil-fuelled armed forces.

What happens when a community tries to live beyond its resources? The best-known if somewhat confused approach is that of the Judeo-Christian mystic who received from Jesus, via an angel, the Revelation of St John the Divine, the last book in most versions of the Bible. The author is unlikely to have been the Apostle John, whose version of the life and preaching of Jesus is in a plain, workmanlike style; the only similarity is that the two Johns write in Greek. St John the Divine's revelation is suffused with the numerology of the Babylonians, common to much of the Middle East, with the prime number seven figuring in visions of seven seals, seven veils and so on whose meaning, if any, has largely been lost. However, discussing the limits imposed on a small, newly agricultural people, the ancient Hebrews, struggling to survive natural and human-made disasters, the prophet sees these personified as four horsemen, as ridden by nomadic warriors, the scourge of farmers everywhere.. Although the symbolism is obscure, these are generally identified as War, Famine, Pestilence and the appropriately pale rider, Death. The expression is cloudy but the ideas behind it are clear enough, and highly topical.

War. The Red Rider is already abroad, according to British Defence Secretary John Reid, and he has resources on his agenda. Climate change, Reid warns, is likely to make "scarce resources, clean water and viable agricultural land even scarcer", which "will make the emergence of violent conflict more rather than less likely". Indeed, considering that three-quarters of the world's known oil reserves, the key to global economics, war-making and elections, are in the Middle East, it would be an odd coincidence if the wars and threats of wars of that region had nothing to do with oil. It is notable that Reid does not mention population pressure as the root reason for these scarcities, presumably because he wants people to vote for his party.

If another species tried to take over our planet, or even interfere with our plans, we would, without hesitation, order a cull, as we have done with barking cows and snuffling hens, not to forget uppity crocodiles and kangaroos surplus to our pets' requirements. When it comes to a human cull, or intra-species conflict, we draw back, and rightly so. Even such renowned cull-artists as Genghis Khan, who made mountains of skulls outside defiant cities, was more concerned with intimidation than extermination: while the Mongol ruler was all for a single world government (his), like a modern investor looking for economic growth, he wanted to encourage production so that he could enlarge his own cut. War between cullers and cullees is a distinct possibility, if likely to be called by some other name.

Human purpose and destiny are much more the province of religion than science, and wars of religion are notoriously bloodthirsty, free of any pity or moral restraint. The Austrian economist Ludwig von Mises, "the bourgeois Marx", identified the invisible hand that ensures that our share dealings serve a useful social purpose as the Hand of God, with any small mismatches being safely left to private charity, but this would horrify a Hindu (which hand, of which god?) and baffle a Buddhist, with no god at all, and living in an external world that doesn't really exist.

It is true, if not exactly reassuring, that religious wars have historically led populations to decline, at least temporarily. The wars of religion of early modern times reduced Germans from thirty million to ten million, delaying those formidable folks' entry into the scramble for empire until unfashionably late in the nineteenth century. When the Western Roman Empire collapsed under the pressure of Germanic barbarians, many posing as good Christians, the ensuing Dark Ages saw one of the only two falls recorded in world history, from 257 million in AD200 to 208 million in AD600. But by AD1000, with the feudal system operational, the world population had almost recovered, to 253 million. Despite the best the Red Rider could do, we have generally been in favour of life, at least for ourselves, and only in the past few decades has anyone felt uneasy about where this might be leading. Adam Smith raised no enlightened eyebrows when he wrote: "The most decisive mark of the prosperity of any country is the increase in the number of its inhabitants" and the same, he presumed, applied to the collectivity, the world. There will never be a shortage of plausible causes to fight over. We can even imagine war between environmentalists and neo-liberals: "Kill those who are killing our Earth" is a slogan observed on recent ultra-Green websites. The success of the Red Rider may turn more on the destructive power of our weapons rather than our skill in inventing causes to fight over.

He must, however, have taken heart from recent developments in China and India, the two tsunamis of humanity threatening to swamp Lifeboat Earth. Despite years of civil war, famine, foreign interference and opium smoking, the population of China reached 500 million in 1948, just before Mao Zedong, one-time chairman of the library committee of Beijing University, became Chairman Mao of his newly proclaimed People's Republic of China. Echoing Adam Smith, Mao saw only good in this mass of mostly illiterate farmers, declaring, "China's vast population should be viewed as a positive asset. Even if it should multiply many times, it will be fully able to resolve the problems created by this growth. The solution lies in production … Revolution and production will resolve the problem of feeding the population."

China's population did multiply, more than once and, such is the perversity of politics, has managed to feed itself, but not by means Mao glimpsed in his worst nightmares. After the Soviets broke the American nuclear monopoly in 1949, Mao saw another use for all those millions. China could win a nuclear war by simply having more dazed survivors on deck when the fallout cleared than any rival. Part of his thought was an atom-proof shelter for himself and his entourage dug into a mountain in Shaoshan, his home village, which I inspected (it was damp and gloomy, but functional) in 1978. But that nuclear war never happened and half-hearted efforts were made to interest Chinese in family planning.

Between 1963 and 1973, China's population more than doubled. In the first three of those years, corresponding to the Great Proletarian Cultural

Revolution, the birthrate reached forty per thousand per year, among the highest ever recorded in a human population, while deaths fell to just over seven per thousand, a result achieved by "barefoot doctors" (simple public heath measures), a sparse diet (1,800 calories a day), little booze and cigarettes (too expensive), and healthy, life-prolonging personal transport by foot or bicycle. In 1966, the last family-planning clinics were closed, followed by campaigns to "learn from the people" (shutting China's universities for four years, with no doctors, engineers or, worst of all, teachers qualifying) and to criticise Confucius, Beethoven, the Italian film director Antonioni and other enemies of the people.

The reaction of Chinese farmers was predictable. Too hardheaded to go forth and multiply at the command of Mao, or anyone else, they simply assumed that he was off his rocker – a verdict history will long debate – and took out the immemorial farmer's insurance policy against times of turmoil, many children to work the farm and support them in old age. But, thanks to Mao's "fair (if skimpy) shares" policy, they and their parents survived. By the early 1980s, when the dark nights of Mao had begun to lighten, China had 550 million under the age of twenty-one just about ready to enter their own child-bearing years. Mao's babies are themselves having babies, so that, even with official rewards for only one child per family and heavy penalties for more, China's population at the beginning of 2006 was a staggering 1,306,313,815 and the fertility rate was still 1.72 children per woman, where it should be one, or better yet, under one, so the population continues to grow, if slowly, and may reach 1.5 billion by the 2020s. Only a tenth of China is cultivable, the rest deserts and photogenic mountains. This is half the land per head available to Indians, one-tenth the share of Americans. Now you know why farmers' sons and daughters by the hundreds of millions are invading China's burgeoning industrial cities hoping for a chance to make your shirts and shorts and soon, your cars and trucks, with millions behind them pleading for their jobs at any wage. China tested a nuclear weapon in 1964: Mao's morose calculation may yet shape her leaders' actions in a crunch.

This is likely to involve India, if President George W. Bush's agreement of March 2006 to share nuclear technology "for peaceful purposes" and Prime Minister Howard's not-now-anyway reply to India's request for Australian uranium, despite India's status as an under-the-counter nuclear power, are any guide. Nuclear power can lead to nuclear weapons, as India's own case illustrates. Prime Minister Nehru, mused in 1949, somewhat anachronistically, that India would not have fallen into British "slavery" if it had possessed parity of weapons. India took a drubbing from China in 1982, although no fallout stained the Himalayan snows that failed to separate the combatants, even though India had tested a nuclear device made by stretching "peaceful" technology in 1974. But the Chinese were not headed for Delhi, and the Indians were headed nowhere, except home. Nevertheless their populations have grown exponentially, on similar lines, with India staying a demographic step behind its giant rival.

In 1950, with the disturbances of partition, independence and civil war behind it, India's population was 358 million, China's 555 million. Last year, India's passed the billion – to be exact, 1,085,264,388, still roughly a quarter less than China's – but the other comparative statistics (courtesy of the CIA) paint a different picture, that of leader and led. China's population growth rate is now down to 0.58 per cent per year, India's is nearly triple that, 1.4 per cent per year. China's fertility rate, 1.72 babies per woman, is below the replacement mark of approximately 2.1 babies per woman, India's well above, 2.78 babies per woman. The Indian tortoise, in short, is gaining on the bigger Chinese hare.

India has shown an inefficient and muddled approach to family limitation ever since independence from Britain. In the early days, there was interest in periodic abstinence (the rhythm method) in line with the principles of self-control advocated by the martyred Mahatma Gandhi. It achieved little. Schemes for intra-uterine devices and voluntary sterilisation have foundered on inexpert fitting and advice.

The real reason is not far away. Poverty breeds, and hundreds of millions of poor Indians (a quarter live below the poverty line) took out the farmer's insurance of large families, whose poverty drove them to the growing cities, where they continued to breed big families. The resemblance to Britain's pioneer industrialisation is striking. Ironically, the social conditions that limit families are well known. They are education, particularly for women, which gives them the choice of living useful lives not exclusively devoted to housekeeping and child rearing. China now has a male literacy rate of 95.1 per cent, female 86.5 per cent, while India's rates are male 70.2 per cent and females only 48.3 per cent. On present trends, India is due to overtake China in population by 2030 or so, both nuclear-armed military superpowers courted as allies, feared as rivals in whatever configuration the world has arrived at. What happens then is, as a Hindu might say, in the lap of the gods. Already, however, India's dash for greatness has had unexpected results. India's middle class now approaches 250 million, all keen on cricket, so three-quarters of the circulation of cricket's Baghavad Gita, *Wisden Cricketers' Almanack* is sold there, and India now dominates the world game. Many of the middle class are highly educated, in English, so India has become a world leader in IT technology and offshore service provision – a legacy, like the railways, of the years of British rule.

Pestilence. The White Rider has had a thin time of it lately but his preferred preconditions – human populations bursting at the seams of mushrooming cities, fast global communications, porous borders – are ever present, and the rider's luck may change. As recently as 1970, I spent a week on a cholera train in India evacuating refugees from the war that created Bangladesh, with the train barely slowing to dump emaciated bodies still wearing the waist-cord haughtily indicating their status as (dead) Brahmins. The cure is simple – rest and clean water to drink – but unattainable in the chaos of war, a reminder that bacilli, viruses and other nasties are forever vigilant, waiting for us to drop our defences while we are otherwise engaged. The Black Death (plague) managed to reduce world population from 442 million to 375 million in a series of waves, probably from Asia, that eventually fizzled in the 1600s.

Famine. The Black Rider can appear in many forms. As long as our oil-based fertilisers hold out, we are not likely to run out of enough to fill our bellies, if not always to find ingredients for fine dining, but this presupposes a perfectly functioning growing and distribution system no invisible hand can guarantee. The world economy, for instance, shows a disturbing resemblance to the last days of the Dutch Republic – gigantic, uncollectable debts, ever intensifying international rivalries, deeply entrenched vested interests, a plethora of reform schemes but no agreement even on which problem we seek to solve.

A worldwide slump would certainly clear our skies of sulphate aerosols, one of the "greenhouse gases" (although it is actually a pall of fine, light-absorbing particles), but no one knows where this would lead. Some researchers see us in an era of "global dimming" and clearer skies caused by reduction in industrial activity which would accelerate global warming, raise sea levels, flood low-lying land and further disrupt our overloaded system. One-crop failure, like the Irish potato blight, could reappear.

Water consumption per head tracks fossil fuel use, and drinkable water too, shows signs of running out all over the world.

Death. The sickly pale green rider is the joker in the pack, the one thing we can expect with real confidence – the unexpected. Death has many options: tsunamis, volcanos, sunspots, asteroid strikes, the One We Never Thought Of.

We could also, of course, be lucky: no one knows what caused the mini Ice Age of the 1700s, and we could always have another, but this would simply postpone the day when the lifeboat will be obviously full and we (to our credit, in our own human eyes) shrink from pulling up the ladder, Jack, while hoping no one else pulls it up on us.

Here the Dutch may still offer a good example. When the water started to rise, they applied small, local palliatives, a dike here and a pump there, rather than grand schemes, and a Dutch firm is still thinking along similarly modest lines, offering a tethered houseboat that simply rises to ride out storm surges and falls back when terra becomes a little firma. What we need to rethink is the economic system the Dutch stumbled across, and failed at themselves. We have Tasman and his ancient island as a reminder and we can thank his lack of curiosity that Melbourne is not called Nieuw Rotterdam and this is not entirely written in Dutch. Nothing can be achieved as long as, under the name of "economic growth", we insist on divvying up every saving in energy we make into shares and selling them to the biggest borrower, thus getting deeper and deeper into debt to our uncertain future.. This is not a moral question: the market system is a means to an end, not an end in itself, and we need another, as yet undiscovered, that will serve us all better. As the economic historian Josef Schumpeter once remarked, stationary feudalism was an historical entity, stationary socialism an historic possibility, but stationary capitalism an historical contradiction in terms, it must continue to expand endlessly, or collapse.

A small start might be a new name, perhaps with an Australian ring to remind us that we are the emitter-in-chief's deputy out our way and, as the world's biggest coal exporter, may well be doing better than the boss. So recent is our discovery of Lifeboat Earth that we have no name for it that doesn't mean something else. Earth is what we put in flowerpots, world means the totality of things in general, monde means a lot of people and so on. James Lovelock now thinks that his coinage Gaia, suggested by his writer neighbour William Golding, was a bit airy-fairy and hard to pronounce, even by literate Chinese ladies. How about Emoh Ruo? This would translate neatly as Suon Zech and Tlew Resnu and, with a bit of jiggling, into !Xosa, Afrikaans (Siuh Sno?) and all the Asian tongues.

Emoh Ruo is an idea whose time has come. We can only hope it has not already gone again. ■

References and other supporting material are available at www.griffith.edu.au/griffithreview.

Murray Sayle is a journalist. His essay 'God wills it' was published in *Griffith REVIEW 7: The Lure of Fundamentalism*, and 'Even further north' in *Griffith REVIEW 9: Up North.*

Debate: **Fair-weather friends?**

Author: **Robyn Williams**

I like Bob Carter. Even in a kilt. He has that baritone warmth that men share when they assume they're united against the Philistines.

I first met Professor Carter in 1998 at James Cook University in Townsville where we recorded an interview outside the old staff club. I like that spot because it is in the open, under the tropical sky and interviews have to be short (under ten minutes) because after that the mosquitoes bite like piranhas.

We talked about Bob's part in the deep-ocean drilling scheme carried out by ship in various parts of the world. The cores produced are analysed chemically and offer time capsules going back centuries and histories of past climate change. He thought the project to be valuable. I agreed, for what that's worth, and the interview duly went to air.

Australia then withdrew from the project and Bob's involvement went on hold. There were two consequences: the first, apparently, was that Carter now had time on his hands; the second, I inferred, was that he was not at all pleased with whichever authorities had cut off the funds.

It is from about that time, four years ago, that many of us began to receive helpful items from Carter, clearly meant for publication, most knocking the orthodoxy, the bleak line on global warming. The first, a scripted talk, I duly put to air. Then a similar piece turned up in *The Australian* newspaper; then he was on *Counterpoint*, ABC Radio National, twice, all with the same position.

This was becoming not so much the availability of a helpful boffin, more pressing a line. His interpretation of the uncertainties went, shall we say, rather further than most weather researchers allowed. I decided to dig deeper and discovered that Professor Bob Carter, geologist from Townsville, was a vocal member of the Institute for Public Affairs (IPA). Fair enough. But the effect of this kind of debating is more like politics than science. First, it presents the issue as one-side-or-the-other, for-or-against. Second, it accuses the other side of what it itself, a lobby group, is trying to do: push ideology.

I decided to put all this on the record and invited Bob Carter into the studio for a chat. I had already knocked back his latest encyclical on climate, connected to the G8 summit, on the dual excuse that I had no spare air time (true) and that I had already arranged a piece linked to G8 based on new, peer-reviewed research, published in a leading journal. Evidence.

We chatted at length, discussed the demise of Australia's involvement in the drilling scheme and then moved to his unequivocal opposition to worries about climate change. They amounted, essentially, to two concerns: the first, his claim that there is "no theory of climate"; the second, that the Intergovernmental Panel on Climate Change (IPCC) and its adherents are either naive at best, or green "religionists" at worst. When I quietly pointed out that I often go to labs like the renowned Scripps Institution of Oceanography in La Jolla, California, where they first discovered the carbon dioxide increase fifty years ago and meet scores of renowned scientists who have made worrying findings on climate ranging from deep-ocean warming and plankton death to atmospheric analysis and bird counts, Bob was unimpressed.

Then he told me that the upper atmosphere is actually getting colder. I had heard this before. It was a weird paradox none of us could fathom. We parted amicably, as usual.

The following day he sent me an email. Would I cut the final remarks? Two papers in the journal *Science* had just exposed the "cooling" story as false. Bob was too good a scientist to let a flagrantly misleading fact be broadcast. But I reckon he was too committed to his position to have any qualms about the rest of his unabashedly sceptical remarks.

That is what's so extraordinary about this "debate". Any reasonable person would say: "This is a complex matter, which, if true, has devastating consequences and must be taken seriously. So far, the evidence, on many fronts, is worrying. There is uncertainty elsewhere. On balance, I am sixty-two per cent convinced the scientific authorities are right and action should be taken." Choose your own percentage. It is likely, if you are sensible, to be in the mid-range. If you opt for either a hundred per cent or zero, I'd think you're cockeyed. Or political in the extreme.

Let's take the four famous furfies of the climate nay-sayers. They are: the troposphere is cooling; it's all in the solar cycle; the glaciers are expanding; and it's a matter of water vapour not CO_2 and methane.

The troposphere is at the top of the surprisingly thin veneer of atmospheric gases covering the Earth. Its temperature has been measured by satellites over the years and, in doing so, incorporated an adjustment for direct solar heating of the apparatus. As the instrument circles the planet, day and night, it is warmed up during the day. This must be distinguished from measurement of the air itself. It was this adjustment that the *Science* papers had shown was wrong. The troposphere is getting *hotter.*

What about the solar cycle? This is about eleven years in duration and its effect on warming – rather than any human cause – has been the line taken by Willie Soon and Sallie Baliunas. I interviewed them both in their office at the Harvard-Smithsonian Center for Astrophysics and they insisted: "The climate of the twentieth century is neither unusual nor extreme."

Chris Mooney also quotes this line in his magisterial book *The Republican War on Science* (Basic Books, 2005), in which he says that Soon feels free to dismiss great swathes of other scientists' findings "based on tree rings, ice cores, corals and other sources". Mooney also notes that Soon and Baliunas received support from the George C. Marshall Institute which is funded, in part, by Exxon Mobil and that their key rebuttal of the climate-change orthodoxy was published under dubious circumstances involving the resignation of a journal editor in protest.

And what about the sun? Well, on October 1, 2005, *New Scientist* quoted the work of Chris Turney from the University of Wollongong, whose research was published in *The Journal of Quaternary Science.* Headed "Climate doesn't swing to the rhythm of the sun", it reported that Turney's studies of peat bogs in Ireland show "peaks in solar activity do not coincide with peaks in warmer conditions".

Which brings us to glaciers. The following stunning statement was made by botanist David Bellamy, writing to *New Scientist* on April 16, 2005: "Many of the world's glaciers," he announced, "are not shrinking but in fact are growing ... 555 of the 625 glaciers under observation by the World Glacier Monitoring Service in Zurich, Switzerland, have been growing since 1980."

George Monbiot, a friend of mine who lives in Oxford, couldn't believe his eyes. He therefore phoned the glacier monitors in Zurich and received the following reply. "This is complete bullshit. Despite his [Bellamy's] scientific reputation, he makes all the mistakes that are possible." How come? Monbiot tracked the source of the snafu: a paper quoted by an anti-climate-change website (no such paper exists) plus an admitted "typing error" by Bellamy (555 instead of 55%). Bellamy's reputation is now in tatters – but the misleading factoid rides on. Three down, one to go.

So, finally, to water vapour. There is no doubt that water is a greenhouse gas and that it exceeds carbon dioxide as a constituent of the atmosphere. There is also no doubt that the main advocate of this theory, Professor Richard Lindzen of MIT, is a respected scientist and member of the American National Academy of Sciences. I interviewed Lindzen in Boston and was impressed by his assurance as well as his cheerful chain-smoking and delight in being contrary. He is known to dispute links between cigarettes and lung cancer.

But what is he really saying about IPCC and the near-universal concern about climate? Does he dispute the stance taken by his own academy in warning the world? Mooney quotes Lindzen as saying he has no essential dispute with the academy's line and is merely critical of the IPCC's summary for policy-makers. The academy gave "an okay summary of what's gone on in the field, read in total" conceded Lindzen. Water vapour is no doubt important in the equation but few use it to discount the entire mountain of evidence from other sources.

I am a science journalist and I report evidence as released in journals where truth is paramount. I may also seek opinion to place some of the findings in context, but it is my job to distinguish between the two and so try to avoid confusing the public. When I read some commentators here and abroad I am overwhelmed by their sudden omniscience about scientific research and also their blithe confidence that they can master vast amounts of information normally outside their province.

So it was in *The Australian* when Christopher Pearson took on Tim Flannery and his book *The Weather Makers* (Text, 2005). "One of Radio National's world view's more disconcerting features is its enthralment to apocalyptic science. Sometimes, listening to Fran Kelly discussing the latest portents of global warming, I can almost hear the polar ice melt and rising waters lapping at the front doorstep. No hint of licensed scepticism, no inkling that this may be just one more millennial fantasy, is allowed to obtrude. This is not a radio show. It's a sacramental observance for true believers." As for Flannery, "he's more shaman than showman, a folk mystic and prophet for the New Age remnant".

And who is wheeled in as a counter-authority? Why, my old friend Bob Carter (no kilt this time).

It reminds me of the "debates" on smoking in the 1950s and 1960s, or the flak we took in reporting on asbestos in the 1970s, or lead in petrol in the 1980s. Give "both sides" was the instruction (asbestos doesn't kill? Lead is good for children?) and on TV we saw one versus the other, as if it were our old friend with two heads, 50/50.

Well, it isn't. There are some predictable columnists who will rail as sceptics of climate change. There are some retired scientists, often geologists like Bob, who will front up in a trice, to run their familiar party pieces; and there are one or two, not many, researchers in the field with some doubts about detail, if not the whole picture..

There are no certainties in science. Most of us are as happy about the prospect of human-induced climate change as we are about the prospect of disembowelment. We would rather it disappeared – like Y2K bugs. But nature is doing something out there, whether we like it or not. If it is drastic, we need to know.

Meanwhile, Dr Barrie Pittock of CSIRO Atmospheric Research in Victoria has just spent months analysing sixty-five major papers on climate, ranging from permafrost melting and global dimming to tropical cyclone incidence and ocean circulation. His conclusion? Scientists look like they have "consciously or unconsciously *downplayed*" the problem. My italics. Our worry. ■

Robyn Williams presents *The Science Show* on ABC Radio National. He has developed his essay 'God's only excuse' published in *Griffith REVIEW 8: People Like Us,* Winter 2005 into a book, *Unintelligent Design,* to be published by Allen & Unwin this year.

Reportage: Knocking on the door

Reporter: James Woodford

Image: Ron Guy / 'Changing Climate' / oil on linen 61 x 71cm / Courtesy of the artist and CUSP Gallery.

Knocking on the door

My home is in a small community on the New South Wales far south coast and when New Year's Day 2006 dawned, the prediction was for a stinker. There was even an expectation the day might be the hottest on record, almost certainly the hottest-ever January 1. Friends were staying and, because we had been up late the night before, everyone was slow to head for the beach. The sun there was as hot as a bluebottle's lash, making twenty-one-degree water feel chilled. A parchment westerly, searching for a bushfire, also blew. The wind and the sun made the sea feel like a haven and we lounged in a rock pool, more senseless hippos than hung-over humans. But consciousness can be a curse and we all knew, even in the rock pool that we were being saturated in ultraviolet radiation.

Back home we put on a brave face. One visitor, a baker who lives on a sheep farm in northern Victoria, declared: "C'mon, guys. This isn't hot." An hour later he was lying paralysed on the concrete-slab floor, a tea towel soaked in icy water over his head and face, moaning like a victim of acute appendicitis. Beyond the shadow of the veranda the sun burned white and the westerly came in through every crack and window. We all agreed it was the hottest we had ever felt, anywhere. It was dangerous heat, the kind of weather that could make people sick or even kill them. The sheep-farming baker could barely talk. A neighbour called to say that in the shade his thermometer read 46.6 degrees – the high end of a series of stunning local amateur temperature recordings, all over 43.9.

Late in the afternoon, a thumping southerly dropped the temperature by twenty degrees and the hot day became the stuff of legends.

In Sydney, the temperature had only hit the second-highest mark ever – a soaring 44.2 degrees – more than one degree short of the record of 45.3 set on January 14, 1939. This fact is critical because Australia has always had hot days. The point is though, if the forecasts are right, the records will continue to tumble. New Year's Day will have been the hottest that anyone under sixty-seven and living in coastal New South Wales will ever have experienced. My bet is that I won't need to live for another sixty-seven years to experience another day like it.

As a prelude to the boiling beginning to 2006, the day before brought down the curtain on Australia's hottest year since records have been kept. The average temperature, across the vast continent, was 22.89 degrees – 1.09 degrees above the 1961 to 1990 average. It was the second time the hottest-year record had been broken in seven years. It may not sound like a huge increase but as the Australian Bureau of Meteorology noted: "While these temperature departures may seem relatively small, a one degree Celsius increase is equivalent to many southern Australian towns shifting northward by about a hundred kilometres ... The 2005 record is yet another sign that our climate is changing. Since 1979, all but four years have been warmer than average in Australia." On the entire continent, as measured by a hundred weather-observation stations, only one region was cooler than the average – the coastal strip of Western Australia from Carnarvon to Cape Leeuwin.

And it was not just Australia. According to the World Meteorological Organisation, the global mean temperature for 2005 was nearly half a degree above normal.

In the days after New Year's Day, people spoke of things the heat had killed – native saplings on the edge of the road, garden plants. One awed resident said even his waterlilies had begun to expire, apparently unable to get enough water through their fleshy leaves. None of those things, however, was a surprise considering the extreme temperature.

A fortnight later I was at the beach again and noticed many of the coastal wattles carpeting the dunes were dead or dying. The local parks service often sprays the area for a weed, bitou bush. I telephoned local parks-service area manager Preston Cope to find out whether the plants had been poisoned, but was told herbicide was only used in winter. He told me beach vegetation had been killed throughout the district. He had no proof that the scorching heat was responsible, but the hot day seemed the only explanation. As an aside, Cope told me what had happened to a different ecosystem near to him. In a single day, the fronds of all his tree ferns had been burnt and killed. His bird's-nest ferns, too, had been fried. Maybe both would grow back in time but what if there were another hot day this summer or a record-breaker next summer? How many times will such species attempt to carry on in a world where extreme becomes average? It was a turning-point moment. Scientists have long said that in a hotter world many ecosystems – whole communities of wildlife – face ruin.

And there it was: an ecosystem in my own backyard bombed by a single day of extraordinary heat. A landscape I love, know and take for granted could be annihilated. All it would take would be a run of forty-six-degree days for my beach to be altered dramatically. We are no longer happy to simply remove individual species. Human destructiveness is on the brink of an ambitious new low – the capacity to inflict the extinction of views, landscapes, snow-capped mountains, ice sheets and coastlines. And not slowly, either. If the dead wattles are anything to go by, in a climate-change world we may not need time-lapse photography to watch a forest wither.

Part of Cope's patch is a beautiful, small island called Montague. It is nine kilometres offshore and always a blissfully cool place. It has also traditionally been seen as the northern boundary for life evolved in the Southern Ocean. Before he hung up the telephone, Cope told me that on New Year's Day the temperature at Montague had hit an astonishing thirty-nine degrees, the highest recorded.

Montague Island looms large from the little coastal town of Bermagui. Bull kelp, a cold-water species forming famous underwater forests, grew there until the 1970s. Today the northern range of bull kelp has moved to Tathra, seventy-four kilometres down the coast.

Seaweed expert at the Royal Botanic Gardens in Sydney, Dr Alan Millar, says the range of bull kelp is contracting southwards at a rate of almost two kilometres a year. "It won't be long before it falls off the edge of the continental shelf and that will be it," Millar told me.

In Tasmania, giant kelp is also disappearing. In 1964, there were 1.4 million square metres of the seaweed. In 2001, there were 700,000 square metres. Another seaweed, ecklonia, which once grew as far north as Caloudra in Queensland, is now found only as far up the coast as Byron Bay. "These organisms are slowly being cooked," Millar says.

There is, however, a converse. "We are finding species only known from the far north of the coast that we are now picking up way south. Tropical species appear to be on the move south."

But life is not just being cooked at the cold end of the continent. A Great Barrier Reef calamity is unfolding; this generation may be the last to see the reef in its full glory. Until recently, I was relatively sceptical about the potential impact of climate change. My view was that if we can't predict the weather a week in advance, how can anyone possibly say what will happen in a century? Instead, to me, the issue was one of waste – regardless of consequences, it is wrong for a few generations to squander a planet's non-renewable energy resources.

I still believe it is not possible to say what the human experience will be a hundred years hence but a trip to Far North Queensland in 2004 convinced me we are in store for some ecological insanity. At very short notice, I was sent on an assignment to cover the environmental problems facing the Great Barrier Reef. I arranged to join Australian Institute of Marine Science researcher Dr Terry Done who was on board the research ship MV Franklin. After a soaking journey in an inflatable, I made a mid-ocean handover from a tourist boat, which had earlier that day taken me to John Brewer Reef, seventy-five kilometres offshore from Townsville, to the deck of the Franklin.

Once on board, there was no time to lose. Our destination was Myrmidon Reef, another seventy kilometres out to sea – one of the most remote and wild spots anywhere in the marine park. The prediction was for a severe southerly front to sweep through the following day, which meant the divers needed to be in the water soon after dawn.

Myrmidon is regarded as completely free of the pressures affecting other parts of the World Heritage Area. Fishing is banned, tourists are few, direct pollution from mainland run-off is non-existent, as are crown-of-thorns starfish. It is inside a green zone, with the highest possible level of protection.

Among twenty-eight Great Barrier Reef study sites, Done has two at Myrmidon and he has been visiting them for twenty-five years. The reef there is his favourite. In 1982, he noticed a tiny patch of coral had turned white. At that time he had no idea the bleaching he had observed was the prelude to mass disaster caused by heat. In 2002, during a vast pooling of warm water in the region, he noticed the reef was stressed but felt confident Myrmidon would recover.

Two years later, as we prepared to submerge, Done's hopes were shattered. "Enjoy the moonscape," he groaned, after quickly looking beneath the waves. The view that greeted us was reminiscent of a bombed city. Bleaching occurs when water warms to a temperature that forces the colourful algae living symbiotically within the corals to jump ship. If heating is not too prolonged then the algae will recolonise. If not, the reef dies. The first impression at Myrmidon was that a sepia filter had been thrown over the seascape – dead corals were covered in brown algae. There were few fish. Instead of being an exhilarating, awe-inspiring dive, everyone rolled back into the inflatable stunned into silence. That day we also dived on outlying bommies off the main reef and they seemed relatively unscathed, but anywhere the warm eddies had settled, especially inside the fringing reef, there was death on a mind-boggling scale. In 1998, forty-two per cent of the Great Barrier Reef's corals were bleached to some extent, nearly a fifth severely. Four years later, fifty-four per cent of the reefs were bleached.

Done told me a one-degree increase in sea surface temperature would raise the incidence of reef bleaching to eighty-two per cent, two degrees to ninety-seven per cent, three degrees and the Great Barrier Reef as we know it would be finished.

This trip made me realise climate change makes a mockery of humanity's meagre environmental efforts. If a coral reef on the very edge of the continental shelf, within the boundaries of a marine park, is not safe, then other places closer to the impacts of our species are in real trouble. Climate change has the potential to undo almost every good conservation initiative ever undertaken. Every national park, landcare planting, water-saving initiative and river rehabilitation scheme is facing the very real risk of being struck from existence. The Great Barrier Reef story is being repeated across the continent. Kakadu National Park faces disaster from increasing salt intrusion into freshwater wetlands and the Wet Tropics World Heritage rainforests are doomed under even moderate climate-change scenarios.

Our efforts to date are nowhere near the sacrifice needed to bring carbon dioxide levels down. We have left our appeasement efforts so late that even an emergency shutdown of all the polluting pillars of society would mean climate change could take decades, even centuries, to turn around. And life may never return to a natural course.

As I flew back home, the jet drew beside the biggest thundercloud I had ever seen. It was whiter than snow and its face was so sharp I felt as though we were flying just metres from the wall of a giant iceberg. Never before had the atmosphere looked so malevolent. I was convinced: talk of losing the Great Barrier Reef is no pinkie-greenie-communist conspiracy. I will be surprised if the reef in its current glory is still there for the centenary of Done's survey.

The ecosystems that will show the bruises from climate change's violence first are those that live in a narrow band of survival – ocean life, mountain-top communities and cities.

On land, one of the clearest pieces of evidence that something big is amiss is that nature's timetable has become as messed up as a badly run public transport system. Migrating birds and seasonal insects like butterflies are arriving earlier and later. Flowers have reset their clocks and some are blooming at times that observers can no longer predict. For creatures dependent on each other, which eat each other or fertilise each other, such changes may spell disaster.

Dr Lynda Chambers is a senior scientist in the Bureau of Meteorology's Climate Forecasting Group. "We are finding that a lot of amateur naturalists are observing changes in flowering, changes of arrival of birds, changes in what birds or plants are found in their gardens and a lot of these changes seem to be related to changes in climate," Chambers told me.

In the Dandenong Ranges near Melbourne, one amateur phenologist has collected twenty-two years of data and her results are most striking because all the talk is what will happen when climate change comes. Listening to Chambers calmly discuss her research, I realise it is clear climate change has been operating behind the lines of our society for years. "Over the twenty-two years of data, for the flowers that flower earlier there's a thirteen- or fourteen-day difference – that's averaged over all the plants. For the species that flower later, they are now doing so by about twenty days over the twenty-two years. We are still at an alert rather than an alarm stage. But it may be only a year or two's time before we hit the alarm button."

I wondered whether these shifts in flowering had happened suddenly or over time. "Gradually," she said.

Climate change is the Mormon of environmental problems – eventually it will knock on everybody's front door. In the past two years, it has been revealed how perilously close Australia's biggest cities are to running out of water. Perth's rainfall has been below average for decades. Even a dam like Warragamba, with the potential to hold two Sydney Harbours of drinking water, has proven insufficient for one of the world's thirstiest cities. Goulburn in NSW and Toowoomba in Queensland are major regional centres whose leaders are thinking hard about sustainability as their dams run dry. Political leaders may escape the sound of the last gurgle from the reservoir in this cycle, but eventually the prediction by author and scientist Dr Tim Flannery of a major Australian city being rendered a ghost town by climate change could come true.

This time the dams may well fill again. However, what seems even more certain is that Sydney, Melbourne, Perth, Brisbane and Adelaide will have to live with chronic water shortages as climate change's grip tightens. Over the past five years, Sydney's dam water levels have dropped to within a single year of completely running out of water. It is nearly eight years since Warragamba Dam was last full – a scary statistic considering a mere week of heavy rain across the catchment is enough to fill the reservoir. The social and economic implications of a capital city running out of water are unthinkable. Imagine Sydney, Melbourne or Perth running out of water. No wonder the Prime Minister has appointed a parliamentary secretary with special responsibility for water, and utilities ministers around the nation are among the most avid weather watchers. New dams, desalination plants, recycling and stealing water from catchments beyond a city's region are all short-term fixes.

Climate change is like today only more so – bigger hail, stronger winds, more rain, longer drier droughts, hotter, colder and dustier. Scarier. While many of the tales of violence in New Orleans have turned out to be exaggerated, and climate change alone was not responsible for the devastation, I would not want to live in a city without a reliable water supply.

In Australia, climate change will hit the wealthy, especially those who live in beachfront or harbourside homes. Late in January 2006, the world's premier science journal, *Nature*, reported the findings of Australian researchers from CSIRO's Marine and Atmospheric Research, Hobart. The team found that by 2100, based on current trends, sea level will have risen thirty-one centimetres. "This will push back typical beach shorelines by around 300 metres," *Nature* reported on January 19, 2006. As most Australians live near the coast, few will be unaffected. It seems an extreme figure and obviously depends on the slope of the land, yet even a three-metre retreat would lay to waste billions of dollars of infrastructure and real estate. It is easy to imagine the environmental damage that will be done to shorelines, by seawalls and other engineering in an effort to protect the homes of wealthy Australians.

Director of the Coastal Studies Unit at the University of Sydney, Professor Andrew Short, says the expense of protecting and moving infrastructure will be extraordinary. He points to structures like Sydney Airport, which is practically at sea level. Eventually, however, every wharf and jetty in the nation will have to be adjusted or moved. "Saltwater will penetrate further inland and all the mangroves and seagrasses will also have to shift. If you get an increase in cyclones, as predicted by climate-change models, you get more winds from the east and north-east and the beaches will rotate," he said. In other words, sand that is normally pushed to the northern end of beaches by southerly systems will instead be driven to the southern end of beaches. Such a dramatic change in wave and wind regimes will have huge impacts on beach erosion.

This is just one of the many changes that are likely to occur, according to climate scientists. In essence, the CSIRO predicts an average temperature increase of between 0.4 and two degrees by 2030 and up to six degrees by 2070. There will be more heatwaves and fewer frosts and more frequent El Niño-driven droughts. The big drought of 2002–03 saw farm output fall by $3 billion. While rainfall is predicted to continue to decline in southern Australia, it will increase across the tropics. A warming of a mere 0.3 degrees will see the area of snow cover in alpine regions contract by eighteen per cent. As habitat is lost, biodiversity is expected to crash and some introduced pests, such as the cane toad, may benefit from soaring temperatures. Scientists also predict an increase in extreme weather events such as the Sydney hailstorm of 1999. That storm dumped 500,000 tonnes of ice on the city, much of it the size of tennis balls, and caused $1.7 billion of damage. As the insurance industry is now warning, a twenty-five per cent increase in wind gusts will lead to a 650 per cent increase in claims. Even for those who like warm weather, the coming century is unlikely to be an enjoyable one for humans. Direct health impacts of warming will include higher incidence of heatstroke and skin cancer. In addition, diseases like malaria are likely to begin to march into previously temperate zones.

Charles Sturt University's Professsor Nick Klomp says most people do not believe a temperature change of a couple of degrees could make much of a difference. "Actually, it makes a heap of difference. Life is on a knife edge and often there is nowhere for it to go. If a species tries to move it hits water, mountains and towns so there are limits to where species can move."

Klomp warns that people need to be careful of being lulled into a false sense of security that the changes are not yet dramatic or apparent to any but the more observant. "Some things change gradually and some things change at a critical level. The best example of a critical-phase change is water melting. Water stays frozen through a huge temperature range but once the temperature hits zero, nothing will stop it from melting." It is such critical-phase changes that are potentially the most frightening of climate change's impacts. Most of these possible phase changes are not yet well understood.

What is understood is that these changes will affect the way we live, our ability to engage with the environment and the nature of the physical world around us – its topography, animals and plants. One of the most ubiquitous Australians is the gum tree, which belongs to a group that consists of 800 different species. Dr Lesley Hughes of Macquarie University's School of Biological Sciences has found that about a quarter of all eucalypts live in an extremely narrow climate range – less than one degree Celsius. With predictions of temperatures increasing by two or three degrees, what will happen to these trees?

Climate change is not going to be our children's or grandchildren's problem. It will be ours. It is ours. My wife, Prue, and I realised several years ago that living in Sydney in a home of our own, with space for children and dogs, was beyond our reach unless we were prepared to spend our entire lives working. We did, however, own land on the south coast and after nearly a decade of wanting to escape the city, we are now living there. Our fifty hectares used to be part of a much bigger farm beside a coastal lake, close to the ocean and looking back on the escarpment of the Great Dividing Range. Nearly three-fifths of our land is a fenced-off forest, untouched and little visited for perhaps half a century.

All around us, urbanisation is taking hold and we knew that if we did not act, our land too, would one day be broken up, the forest cleared and its extraordinary values killed – a death by a thousand cuts. We were determined that we would do everything we could to protect the forest and its wildlife from ourselves, our children and anyone else who might one day own our property.

We placed a voluntary conservation agreement over thirty hectares of eucalypt-dominated forest and now take our obligations under this agreement with the state government very seriously. Most importantly, the forest is legally protected from subdivision. We have been pulling out weeds, planting trees in erosion gullies, encouraging the native species, providing habitat for birds and animals. But it may well be that attempting to protect the forest from being cleared or subdivided is a waste of time. Climate change has many weapons in her arsenal and putting a line on a map and defending it with English property law will not keep her out. Even if a temperature increase doesn't make our forest vulnerable, increased bushfires, weeds and disease may.

On the cleared land, we have begun establishing a native hardwood plantation. It is predominantly made up of gum trees and, while our tree planting is on a relatively small scale, it has consumed a considerable amount of time and money. Hughes' research suggests a betting man wouldn't rate the odds of it ever reaching harvestable age as high. Knowing this, everything I now plant is chosen for its capacity to cope with a large climatic variation.

At the end of the day, electricity consumption is at the heart of why we face such a climate-change mess. The generation of electricity is the biggest contributor to climate change. The fossil fuels that power the turbines are not renewable and the waste they produce is polluting our entire planet.

We did not want to give up on a lifestyle powered by electricity, but we did want to live smarter.

Moving beyond the reach of a major power company is not easy, and last year I discovered just how powerful electricity companies are. The landscape around our home is blighted by powerlines. They are not only ugly but keeping the land beneath them clear of vegetation creates moats dividing ecosystems. The one across our place has cleaved the local forest in two. An aerial photograph of the area reveals a chopping board of powerline cuts. A power easement crosses our property and a line passes within fifty metres of the home site.

We had wanted to rely on solar power but were advised against it. We were warned about the price and perceived technical difficulties. In rural areas, the cost of connecting to the grid is borne by the householder and can often run into tens of thousands of dollars. When we contacted the local monopoly for a quote we were not surprised to be told it would be between ten and fifteen thousand dollars to tap into the closest power pole – the one we would be spending the rest of our lives looking at whatever power source we chose.

"But exactly how much?" I asked

"Well, a formal quote will cost you $500," the officer said.

We decided to start counting our solar savings immediately. I was prepared to live a dim and difficult life rather than pay an outrageous quote to a corporation with a monopoly to make a fortune and destroy the environment.

We planned to use a generator for power to build the frames and part of the roof of the house before installing a solar system. My brief to our local supplier was to install the best system possible for less than the cost of connecting to the pole. I wanted the inside of our building to be as "normal" a home as possible. Subscribing to the sustainable hedonism school of environmentalism, we did not want to send our family back to a Palaeolithic existence. On the day the panels were bolted onto the roof, I asked the supplier about the limitations of the solar set-up. Could we use power tools to finish the building?

"No worries at all. Anything less than 2400 watts should be fine," he replied.

The two builders working with me were sceptical as we plugged in drills and drop saws and continued building. To our amazement the solar system was incredibly robust. The rest of the house was built using power from the sun and, at the end of a full day (even in winter), we struggled to reach night with a battery capacity of less than 99 per cent. During several blackouts we kept working, powered by our own plant consisting of seven desk-sized panels on the roof and a battery box the size of a chest freezer. We bought an energy-efficient fridge from the local appliance store and installed a house full of guilt-free lights.

We began enjoying solar-powered banana smoothies and I started writing solar-powered stories. One night, after a huge thunderstorm swept out of the Wadbilliga Wilderness, taking out mains power throughout the area, we sat and ate dinner, our lights shining and fridge humming. Ours were the only lights in the district that evening.

Our neighbour looked on enviously one week as twenty blackouts threw him into a frenzy of clock and appliance resetting. Yes, there are energy costs in both batteries and panels and it takes years to repay the bill if account is taken only of power bills. But if you add in the price of a blackout preventing me from working, freezers full of spoiled food, not to mention the cost of connecting to the grid, then the debt is quickly repaid. The panels will last decades and the batteries at least fifteen years.

Our home is the same as any other. The power points are identical, it isn't dim or difficult, the maintenance is minimal – once a month I lift the lid of the battery box and check water levels. Best of all, we never get an electricity bill. As Klomp says: "People have grown up thinking they can just push a button and get anything they like. Everyone has now got to have an attitude change. It doesn't mean being uncomfortable. It means a few new tricks."

To me, terrorism is not as scary as the fact that some flowers bloom weeks earlier than they used to or that we don't have a clue what happens when an ocean current breaks down. The only thing scary about al-Qaeda is that they are likely to be the big winners from the social and economic dislocation wreaked by climate change.

Even without global warming, it's a no-brainer that it is wrong for a handful of generations to use almost all the easily obtainable fossil fuels. Even if there were no such thing as climate change, the amount of waste in our society is disgusting – why did the four chocolate biscuits I opened for guests this morning come inside a box, a plastic bag and a plastic container?

Energy bingeing is not only lazy and greedy, there are consequences. Did we really think we could burn fossilised forests and swamps from the age of dinosaurs without consequence? Did we really think that voting for George W. Bush or John Howard was good for anything other than short-term self-interest? It's not the economy, stupid. It's the ecosystem.

Men in suits talk about technical solutions such as carbon sequestration and zero-emission coal-fired power stations. The public only needs to note who is putting such ideas forward to understand they are code for business as usual. Climate change is starting to look like a planetary manoeuvre to expel a species too big for its boots.

My prediction is that the twenty-first century will shatter one of the great and most harmful human delusions. We are not mere observers of the food chain, a separate class of life on Earth. For the first time, we will see that humanity's existence is as fragile as that of any other big charismatic carnivore. ■

James Woodford reports on the environment for the *Sydney Morning Herald*. His most recent book is *Dog Fence* (Text Publishing, 2004).

The Salt Chronicles

Bears no resemblance to the Australian Army Education magazine of the Second World War...

John Kinsella

1. Aloneness

I realise: so often
I write myself alone
as if no one would go there
for its own sake: they might
for science, or surveillance,
reflecting on their loss,
the irony of making more land
less arable, or maybe
tree-planting.

I went there
to enclose in an open-ended
environment, where earth-cracks
from the tearing were gateways
to a journey: social
misunderstanding at school,
competitive sport,
rivalry; centre of the earth.

The formation of a salt crystal
is bridge and timeline to tether
astronomy and forecasting,
to mirror then encrust then paper the bones
of rodents, small marsupials, birds:
the familial breakdown.
Of myself, I was sure:
the tufts of survivor grass,
the resilient spiked trees
stunted and wind-bent,
samphire on the edges and elevations,
marsh grasses slung with Christmas spiders,
stalked by plovers and herons.
In the brine the tumbles of larvae,
crustaceans that shouldn't have survived the salt,
species created in a few seasons
then lost.

2. Salt Wraiths

Salt wraiths leave trails —
salt breaking through clean ground.
Where the salt saturates
they are the white ache
of pillars, arches, sheets:
embedded as insect corpse
or blown seed of wild oats.
At night, they make
a chemical heat.

Of no order,
they connect with nothing but salt
leaching up, or running
underground passages:
above, the owl strikes quick
in its fearlessness,
but flies below the moon
hoping to swoop
out of the wraiths' sonar-
blip, a clash of technologies,
the wraiths emanating
from the ground up,
they take hold
bit by bit,
the owl complies
digesting the mouse,
the night's castings.

X-raying for animal and mineral presence
written stages of sedimentary formation
wraiths imagined they'd known
others — had known song and conversation,
charred criteria of fires; oil lines
blurred in salt buffers, a slush
of samphire crusting
as dryness set dimensions
in rip-up-marks to break it up,
iterations of she-oak
whispers, such small seedlings
to get a hold, but if they do they flourish
and serve to strain; in dispossessing
wraiths might think they displace
but dialogues about the feeding of the river
by salt creeks that will drive out the serpent
are persistent; it's the wraith's indifference
I missed as a child wandering
the first-degree sunburn
and thinking hallucinations
were prophets or ghosts; in the blaze of white
I lost definitions; as a long way north
it's as if focus was made through Bradshaw noticing
or bothering to record when 17 000 years
was a dynamic counterpoint
in itself.

The air is fetid about the gullies' throat:
the rubbish used to throttle
erosion: they batter negatives
against the resistant plates,
these emanations of electrolysis,
afterimages we pick up on
when alone and receptive,
further out in the blanks
it sterilises.

3. Mapping and Companionship

Sketched on graph paper
intended for school, red lines
mark salt seams, blue lines
hard clear water
of gullies and creeks,
green the algal displays
inside their aquariums.
In the drawing out
so some might follow
as nemesis said or echo
in the mirror, or Diana
perved on from afar by binoculars,
or the memory of salt crackle underfoot
the tinnitus that scratches and flutters
like half-formed auditory
hallucinations; fight as much as we did,
my brother and I would go out there in maps
of our ulterior making, and own what cousins
owned by right of family, and own
the fragile nature of the eroding
footholds, lines of wash
from paddocks still yielding
good crops though closer
to granite cap-rock
year in, year out,
marked on bed-heads
by brass shell casings,
sharp lights like the green-gold glow
of navigation markers
dropped by passing aircraft.

4. Contrary

I cannot look at salt
on the crumbling winter roads
of Ohio without it causing
dislocation, a deep disturbance
in what might happen,
a shifting of sensibilities,
a fraught transaction —
even pain.

The salt that hardens arteries,
the salt whose lack has the shearer crippled
on the shed floor, kicking like a wether.

The agistment of salt mines
and the way sweat and blood
dissolve with history: the paranoia
that says underneath it all
must be holes in the text,
Macherey's unconscious urge to
see the oppression of salt licks,
sheep huddled around the drum,
cobalt in their bellies,
manipulating salt taxes,
the market value,
salerium argentium,
the enforced purchase of salt
by children eight years or over,
straddling ant trails
reaching into dead zones
where insects drop
from airspace
and are collected,
collated.

5. Salt Pans at Dampier: Company Semi-Fact Sheet

It takes 18 months
to put the salt
through
its evaporative cycle.
Algae are "contained"
by milkfish
bred to scour the ponds,
to run the mirrors
of sunshine that turn
sight inside out.
Dampier Salt = Rio Tinto 65%
Marubeni 20%
Nissho-Iwai Ltd 10%
Itochu Corporation 4.5%.
They enjoy the nearby
gas deposits, the export of iron ore
through the heaviest tonnage-
capable port in Australia.
The salt ponds = 100 square kilometres.
Magnesium sulphate,
magnesium chloride,
potassium chloride.
World price for bromides
not adequate.
My father managed the workshop
keeping the belly dumping trailers
for the giant salt trucks —
and the Kenworth
prime movers themselves
in good working order.
Management, he wasn't Union,
but respected Union labour
well enough, but "not the blokes who'd
go out at the drop of a hat..."
he tells me this twenty-six
years later — knowing I'd be Union
as the salt drives his blood pressure,
hijacks his sarcasm.

Memoir:
Confessions of a weather nut
Author:
Phil Brown

Image: A severe thunderstorm approaching Surfers Paradise, Queensland / Photographer: Radek Dolecki

Confessions of a weather nut

It was an innocent enough remark. "Looks like rain," he'd said, gesturing toward the bank of clouds looming over the city.

"No way," I said, shaking my head.

"Well, those clouds look pretty threatening to me."

"Those are not rain-bearing clouds," I insisted. "It's a common mistake to assume that because a cloud has a darkish tinge to it that it will result in precipitation but this is not necessarily the case."

"Well, they said showers on the weather report this morning," he went on. He wasn't giving in without a fight. Have you ever met a taxi driver who didn't like to argue?

"They say showers because they have to cover their arses in case there is even the slightest spit," I said, frustrated. "And the media interprets that as rain because no one knows anything about meteorology. But I can assure you that those clouds will not result in any precipitation and that, in fact, it probably won't rain until the day after tomorrow when an inland trough moves across the south-east quarter and out to sea."

He grunted and turned the radio up – John Laws, of course – by way of punishment.

Little did he know that I was no ordinary passenger, however; that I was not the sort of person you could just indulge in idle chitchat with about what goes on in the heavens. For I am a weather nut, a member of a worldwide fraternity for whom the weather holds a fascination that is beyond the ken of others. We are the trainspotters of meteorology, the geeks of the elements.

As a weather nut, I follow, with dutiful dedication, all daily weather reports in the major print and electronic media and I constantly monitor the Australian Bureau of Metrology's website – the BOM as we in the trade like to call it – to keep constantly abreast of matters meteorological. When you hear someone say: "I've just been on the BOM", you'll be able to identify him or her as one of us: an obsessive weirdo – a weather nut.

As far as we are concerned, the intriguing nuances of weather, infinite in variety and constantly evolving, are unparalleled by any other phenomenon in the natural or technological world. Climate change is as alarming for us as it is for anyone else but that alarm is tinged with excitement as we draw closer to a new frontier of weather. *The Day After Tomorrow* may have chilled most of you but we were on the edge of our seats with anticipation as the new ice age ensued. Just a bit of weather, really, and the more weather the better.

As a weather nut, I'm concerned about climate change but there's a part of me that yearns for its extremes. As I wait and watch, I keep abreast of the trends through various weather websites and particularly through the excellent support programs of The Weather Channel on Foxtel. Here I can check rainfall – last twenty-four hours, since 9am, since 3pm – monitor dam levels, check rain forecasts a month ahead and generally track systems as they develop all over the continent and around the globe, for that matter. As soon as my wife leaves the room I change channels like some furtive porn watcher, quickly channel-surfing for the latest conditions and to see if there have been any changes in the forecast in the past hour or so. It's a big wide world of weather out there and if there's none in your own neighbourhood, as there often isn't in my home state of Queensland – beautiful-one-day-perfect-the-next land – there's sure to be something happening somewhere so I can get my vicarious kicks. When some boring high-pressure system hovers over the continent, banishing the colour and movement a weather nut desires, there may at least be a typhoon bearing down on Honshu or even a few tornadoes tearing their way through the heart of America. We understand your pain, dear residents of Kansas, but please let us tarry with you awhile and share the drama of your latest twister. We have no shame, no conscience, when it comes to meteorological matters and, well, we like to watch.

Of course, one isn't born a weather nut. I became one by degrees, inching towards my compulsion over decades, without even my dearest friends and relatives recognising the slide. It began when we moved to the Gold Coast when I was thirteen. My memory of arrival is largely littoral and meteorological. I recall we were staying at a motel in Surfers Paradise until my father sorted out a rental property and I remember sitting on the beach watching the surf. It was, if I recall correctly, a relatively temperate day, say hovering around the late teens (temperate for that part of the world) and a mackerel sky (small cirrocumulus and altocumulus clouds giving a scaly effect, like a mackerel's epidermis) spread above me. The wind was light – say five knots out of the south-west – and a low groundswell pushed in from the south-east, no doubt the result of a loose gradient low slipping down into the Tasman Sea.

After renting a house on the Isle of Capri for a time, we moved into a purpose-built mansion at Cypress Gardens on the Nerang River, a few kilometres inland from Broadbeach. As a student at Miami State High School, I realised that acceptance and even survival required me to become a surfer and, as any devotee will tell you, surfers need to know about weather. We depend on shifting sands and oceans swells and, we hanker for the offshore winds that hold them up as they march to their inevitable doom across the continental shelf. Any keen surfer develops skill as a weather watcher and can feel an impending south-easterly change in their blood, can forecast change in the hooked stratus clouds that precede one, can tell from the tightly coiled isobars on a weather map that there is a groundswell on the way.

As well as surfing, I was a bit of a fisherman in my teens and I also came to know the ways of moon and tide, when the bream would bite or the whiting might nibble. I spent my days studying the sky like some pagan seer, looking for telltale signs in the heavens.

Later, my weather ways fell into disuse for some years as I followed the ways of the flesh, oblivious of the seasons and cycles of the natural world. Enmeshed in work and the sorts of escapism the concrete jungle provides a young man, I became estranged from my former intuitive grasp of the elements. But by the good graces of the goddess Gaia, and with a little help from the surf god Huey, I returned again in the early 1990s when we moved to, of all places, Melbourne. Inspired rather than depressed by the four-seasons-in-one-day realm in which I found myself, my passion for weather returned to me anew. I pored over the weather in the daily newspaper once again, divining meaning from the synoptic charts trying to predict the weekend weather and surf, checking in with the BOM occasionally (you can ring these guys if you want breaking weather info – one weather nut to another) or the guru of weather for surfers, Mike Perry on the Gold Coast. His Surf Alert service tracks weather all over the globe for surfers who need to know.

Most Queenslanders shiver in the southern winter and yearn for the heat and humidity of home but I developed a love of the long, overcast months of the Melbourne winter, the drizzling days and the drama of the southerly busters that cross Port Phillip Bay with such intent. I was in weather heaven.

By the time we moved back to Queensland, I was hooked again and was worshipping the elements with élan, like some druid. At work, I spent my days tracking the various weather systems for my colleagues, warning of impending storms and whether or not they bore hail and damaging winds as well as heavy rain. I was dubbed Punxatawney Phil, after the Pennsylvanian weather-forecasting groundhog of the same name – a great honour. You'll know about Phil (the seer of seers, sage of sages, prognosticator of prognosticators and weather prophet extraordinaire) if you've seen the movie *Groundhog Day*. Each February 2 in Punxatawney, Pennsylvania, United States, tradition dictates that a groundhog is held aloft and if it casts a shadow it is an omen of bad weather to come but if the day is cloudy, and hence shadowless, it is taken as a sign of an early spring. It is an ancient tradition that goes back to the Romans, who used hedgehogs apparently, which just goes to show how long weather nuts have been around.

"Will it rain today, Punxatawney?" a colleague would ask and I would shake my head in disgust.

"No, it will not rain today. Didn't you listen to the weather forecast this morning?" I despair of the ignorance of people when it comes to weather. It takes the impending doom of climate change for them to even notice the weather at all. I marvel at the unconsciousness of my fellow humans who foolishly plan huge public events in storm season and are so surprised when they are washed out. I know photographers who plan elaborate fashion shoots with no regard for the weather. I've heard them making plans and had to intervene.

"You're not really planning a fashion shoot for Friday?" I'd ask.

"Well, yes, the model and stylist are all booked."

"Are you aware that there is a rain event expected?" I say.

"A rain event? What's that?"

"It means it's going to piss down." A rain event is a significant confluence of elements, which result in considerable rainfall – not just a measly five or ten millimetres. We're talking soaking rain, the sort we weather nuts live for.

A rain event is something to look forward to, of course, but it can't beat a storm. Tracking a storm cell on the BOM site, watching it tick across your computer screen like an angry amoeba, is very heaven for any weather nut. The beauty and colour of a storm's topography from the weather satellite … those wonderful blues and greens, the intense yellows or the angry reds at the epicentre, where hail and wind and lightning live. Ah, me, no one but a weather nut understands the emotions this can evoke.

A weather nut likes to share the passion and pass on the craft. (What is that poem by Dylan Thomas? 'In My Craft or Sullen Art?' He could have been talking about weather forecasting just as easily as poetry.) So it was one afternoon this past summer that I felt the urge to begin the succession to my five-year-old son, Hamish.

It was a quiet Sunday afternoon and I was disporting myself on the couch, resplendent in grubby King Gee shorts and little else, watching the cricket. A rivulet of sweat ran down my face, dripped into the thicket of hair on my chest (greying hair, sadly) and proceeded to run down my modest paunch. This signified humidity. I noticed there was little wind outside and, though it had been brilliant and unbearably sunny that morning, a darkening outside suggested clouds at last. Rousing myself, I left the cricket to its own devices, shook off the torpor that settles inside a weatherboard Queenslander on such a day, took my son by the hand, despite his protests that he hadn't finished building his Lego plane, and led him out into the front yard and on to the meagre verge of grass that serves as a buffer between us and the bitumen.

"What are we doing?" he asked quite reasonably. I could have said: "If you can snatch the pebble from my hand, grasshopper, you are ready to leave the temple and go out into the world" but instead I raised a hand, gestured towards the south and said, with gravitas: "Look there, my boy."

Beyond the poincianas and frangipanis, above the railway fence at the end of the street and beyond the hunched city, a bank of cumulus clouds was billowing like a vast, fluffy mountain of cotton wool and shaving cream. In the middle of this mass of cloud, one was well on its way to morphing into that grandest of all collections of droplets, the cumulonimbus, a head of cloud full of lightning and hail. The westernmost edge of this magnificent summer system had already caught up with and swallowed the westering sun, and birds, sensing the coming fury, had already begun winging their way north.

"Look," I said again, challenging the lad to regard the majesty of nature and silently exhorting him to tremble with anticipation before it as I did. It was a grand sight to watch that meteorological symphony working itself into an overture that would soon split the afternoon asunder in a crescendo of driving wind and rain, a quenching deluge that would be brief but decisive and would completely change the complexion of the day.

Afterwards, the evening would be cool, the garden would drip in all its subtropical fecundity and we would sleep easy, relieved of the ennui that had engulfed us.

And as I stood there inspiring wee Hamish to behold the wonder of nature with me and revel in its glory, too, he did look up, shielding his eyes with a hand at first. I thought I detected words of wonder trembling on his lips but he refrained from speech, shrugged his shoulders and looked at me quizzically.

"It's just a storm," he said then, turning and walking back down the path and up the stairs into the house, leaving me stranded in the face of the oncoming tempest. I stood there like Lear out on the heath, forsaken, the storm bearing down. It's lonely, sometimes, when you're a weather nut. ■

Phil Brown in the author of *Travels with my Angst* (UQP, 2004), his new collection of short stories 'Any Guru Will Do' will be published by UQP in 2006. His essay 'Our man up there' was published in *Griffith REVIEW 9: Up North*, Spring 2005.

Reportage:
The gang of six lost in Kyotoland

Reporter:
Graeme Dobell

The icebergs were melting and no one seemed to care. Except the guy who had to clean it up.

Image source: Cartoonstock.com

The gang of six lost in Kyotoland

The rotating vagaries of diplomatic timetables decreed that the United States unveil its climate change trump card on the banks of the Mekong River. The new answer to the danger of rising sea levels went public in the tiny capital of landlocked Laos. The voluntary solution to global warming first saw daylight in a communist state still run by the military.

The Asia-Pacific Partnership on Clean Development and Climate issued its manifesto into the cavernous space of the overgrown hangar that serves as the Laotian Convention Centre. The centre sits amid farmland on the edge of Vientiane. So, a hundred metres away, a lone farmer worked his field with a hoe as Alexander Downer chaired the press conference that launched the partnership.

The key point – almost the only fact in the announcement – was the membership. And that list made the splash. The US had found Australia the easiest of catches, the most enthusiastic of volunteers. It was the other members that were the substance of the headlines – India and China, along with Japan and South Korea. India and China, as the coming economic giants and new great polluters, had signed up to a partnership that had no emission targets and no enforcement mechanism.

The science will be endlessly argued. The new geopolitical fact is stark. In the last decades of the twentieth century, Asia-Pacific institutions could be created with an Asian membership based on Japan, South Korea and the ASEAN states of South-East Asia. In the first decade of the twenty-first century, India and China are indispensable.

The partnership announced itself while tepidly pledging not to undermine the 1997 Kyoto Protocol, the treaty to limit global greenhouse-gas emissions. Kyoto's supporters clothed their contempt for the new partnership in condescension.

The birth notice of the partnership was a terse statement issued from the White House by US President George W. Bush a few hours before the press conference in Vientiane on July 28, 2005. With paternity clearly established, the US stepped back and allowed Australia's foreign minister to chair the announcement.

Downer said that, as the six partners "account for about half of the global GDP, population, energy use and greenhouse emissions, our collaboration can make a significant impact". He pointed to the vision statement that "outlines the core principles and our shared vision".

The key element in the one-page document was that it would be a "non-binding compact" to develop and transfer cost-effective and cleaner technologies. The vision listed seventeen broad areas for possible collaboration. "Possible", mind, and then again there might be other areas they hadn't thought of. The list ran from clean coal to carbon capture, from nuclear power to rural and village energy systems, from home construction to hydropower, and not forgetting the wind and the sun among the renewables.

Downer was the ranking official, the only foreign minister on the podium. Japan's foreign minister was off elsewhere chasing votes for Tokyo's drooping effort to get a permanent seat on the United Nations Security Council. China's foreign minister had already left Vientiane for a more pressing engagement – to offer diplomatic comfort to Burma's military regime.

The US Secretary of State, Condoleezza Rice, hadn't made it to Laos for the annual ASEAN Foreign Ministers' Conference involving more than twenty Asia-Pacific countries. The world's reigning diplomatic superstar doesn't do many remote-country gigs (as Sydney found when it hosted the first full meeting of the partnership in January 2006 without her).

The US representative sitting at the other end of the table from Downer in Vientiane was the US Deputy Secretary of State, Robert Zoellick. He went directly to the line that has become a mantra of the partnership – its non-binding philosophy will complement, not replace, the rule-based targets of Kyoto. But not only will the partnership complement Kyoto, according to Zoellick, it will be better than Kyoto: "One can't just command other parties to do things. You can try, but it's not going to be effective, so you need to try to develop interests and incentives. The US is a member of the global climate convention – the agreement that was done in 1992. We've stated our differences with the Kyoto treaty. So we're committed to trying to address this effort, we just think that there's a better way to do it than the requirements of the Kyoto treaty."

One way of keeping track of the winners and losers in diplomacy is to chart who sits at the table. And as the journalists filed out of the Vientiane conference room, it was clear one country had been snubbed. Diplomats from Canada roamed among the reporters offering an instant response to the creation of the partnership.

The equation was obvious. Canada didn't get new-partner status from Bush because it had refused to join the war in Iraq, turned down a role in the US missile defence program, and some in the Canadian government had been indiscreet in discussing their belief in the "Bush is a buffoon" school of analysis. And Canada was deeply tainted by its support for the Kyoto process. Had Canada followed the lead of the Bush White House and not ratified Kyoto, it could have put the treaty to the sword. Instead, Canada and Japan ratified and opened the way (when Russia eventually joined) for Kyoto to come into force in February 2005.

The modern media rule about applying immediate counter-spin came into play only minutes after Downer wound up the first partnership press conference. On the other side of the Vientiane convention centre, Canada's Foreign Minister, Pierre Pettigrew, strode from the Canadian delegation room to talk to a coven of journalists assembled by his diplomats. It was an exquisite example of praising with faint damns.

Pettigrew measured the members of the new Asia-Pacific partnership against the language of their founding document: "Well, first of all, this is a vision statement. So if the vision acknowledges that there is a problem, this is already progress. Second, I would say that in the vision statement they have acknowledged that this was a complement; it was not meant to replace Kyoto, but it was a complement to Kyoto. So when you want to complement something, you recognise that the real substance is somewhere else. A complement normally is something that adds on to something which is the real thing."

The correspondent from the Australian Broadcasting Corporation thought that an elegantly applied Canadian kick, but went straight to what had been the triumphant underlying note in the Downer–Zoellick press conference. I asked: "Does this, though, deal with the great gap that Kyoto has – this gets China and India to the table in a way that you had not been able to?"

As the microphone swung back to Pettigrew, he summoned up a political line that often runs on both sides of the US–Canada border. When questioning the lack of substance or detail in your opponent's argument, the question to ask is based on an old hamburger ad: "Where's the beef?" The Canadian foreign minister repeated his joy that more countries were preparing to confront the problem of climate change and then swung with gusto: "The words I see in the vision statement are, 'We'll be working on non-binding things'. So I still have to wait for the meat. I mean, I'm pleased with the vision that is there. It is an acknowledgement of this problem we have with climate change. This is an improvement. This is progress, but I am still waiting for the meat."

When looking at Australia's climate-change policy, the problem is not so much locating the beef, but reconciling the contradictions. Laying out the call and counter-call of Australian policy produces a strange maze.

For instance, the Australian Government strongly believes in market solutions – except the market created by Kyoto. An Australia that now puts its faith in business and technology to solve the greenhouse problem has turned its back on a unique new market. In "Kyotoland", you can trade emissions and also get carbon credits for helping developing countries take up clean technology.

Australia ratified the United Nations Framework Convention on Climate in December 1992, one of the first countries to do so. The Howard Government went to the Kyoto conference in 1997, negotiated hard and signed up to what was to be a mandatory target to deal with greenhouse-gas concentrations in the atmosphere. Australia accepted a target of 108 per cent of its 1990 emissions, averaged over the period 2008–12. This eight per cent increase over the 1990 base year was hailed by the Government as a negotiating victory for Australia because other developed countries agreed to reach targets below their 1990 figure.

Australia says it is committed to its Kyoto target, but has decided not to ratify the Kyoto Protocol. From there, the maze gets worse. The Howard Government says it will not ratify Kyoto because it would hurt the economy and cost Australian jobs. But the Government then says Australia is on track to reach the emission targets set by Kyoto (presumably without killing jobs). The Environment Minister, Ian Campbell, said that "while the Australian economy is expected to almost double between 1990 and 2010, our greenhouse-gas emissions are expected to grow by only eight per cent".

Like many factoids in the global-warming debate, Australia's statement that it will meet its Kyoto target is subject to an argument that can only be called heated. (This qualifies as the standard Kyoto quip.) The claim that Australia will meet its Kyoto commitment (saving eighty-five million tonnes of greenhouse-gas emissions a year by 2010) is dismissed by critics as the product of rubbery figures and funny counting, based on "voluntary" reporting by industry.

The Australian maze really hit the haze at the giant UN climate conference in the frozen streets of Montreal in December 2005 (10,000 delegates, 189 countries). Australia's environment minister headed to Canada, proclaiming that the Kyoto regime of mandatory targets was virtually dead because "the cold, hard reality is that the developing nations will not sign up to targets and timetables". But, at the crunch moment, Australia split with the US and helped launch the next stage of Kyoto – negotiations on a new protocol when the current regime expires in 2012.

The US delegation stormed out of the Montreal talks at the penultimate moment and the meeting looked doomed. Strangely, it was America's allies in the Asia-Pacific partnership that seem to have been decisive in getting the Americans back to the table. Even Australia wouldn't follow the walkout. And, more significantly, China said it was ready to take part in talks about what should be done to deal with climate change after 2012. The isolated US moped back to the conference and an agreement was sealed.

Kyotoland was brought into existence with compliance rules and trading of carbon credits across national borders. And those countries already resident in Kyotoland (the developed economies minus the US and Australia) agreed to negotiate on binding emission targets in the second phase beyond 2012.

When the six countries of the Asia-Pacific partnership met for the first ministerial meeting in Sydney a month later, they were not surveying the smoking ruins of Kyoto. Instead, a founding tenet of the charter was that its purpose was "to complement but not replace the Kyoto Protocol". The charter states that the first purpose of the partnership is being to create a "voluntary, non-legally-binding framework for international co-operation". These words are important for the two new giants at the table – China and India – and the other giant denied even observer status in Sydney, the European Union.

Kyotoland is fascinating because it expresses much about how Europe thinks multilateralism should work. Europe breathed life into Kyoto when the US sought to kill off the protocol. The EU's ardent wooing of Moscow (with such carrots as support for Russia's entry into the World Trade Organisation) brought the Kyoto vision into being in February 2005. Russia's enlistment meant the protocol could come into force because it had been ratified by countries covering fifty-five per cent of total 1990 carbon dioxide emissions by developed countries.

Europe, as the ultimate transnational experiment, sees Kyotoland as a new multilateral expression of many of its internal workings. The Australian bureaucracy has spent decades battling Europe over agriculture and the scars explain Canberra's scepticism – even fear – of the European as a behemoth with many heads. Australia has its own version of Kissinger's old question: "If I want to find out what Europe thinks, whose telephone do I call?" The Canberra version today runs: "How do you get Europe to change its mind on anything when all the bargains and concessions have been struck among the twenty-five member states before the EU even starts the negotiation?"

The European Commissioner for External Relations from 1999 to 2004, Chris Patten, expresses Europe's passion for Kyoto when he rails against the US and Australia as energy guzzlers whose "dangerous pitch" to China and India is that the world can confront global warming merely through voluntary action: "We face a common threat; the developed countries have done the most to create it; the rich should bear initially the largest share of responsibility for tackling it. In time, we shall need developing countries to join the effort. That will require persuasion. How do rich countries persuade poor ones to act, if the richest country of all refuses to budge? At this point relative politeness is strained beyond breaking point. US policy is not only selfish but foolish and self-destructive."

Patten, from a different direction, comes finally to the same conundrum confronted by the Asia-Pacific partnership when he concludes: "We all need a Bush conversion on the road to Delhi and Beijing. Unless America is prepared to accept its environmental responsibilities for the future, it is difficult to see how we will ever get India and China to do so."

The twin roads to Delhi and Beijing will complicate Australia's ambition to become the world's largest uranium exporter. When Howard was in Delhi he was pressed on why Australia could sell uranium to China and not to India. And as China's Premier, Wen Jiabao, headed to Canberra in April to witness the signature of the Nuclear Transfer agreement, he gave a mischievous conditional endorsement to the sale of uranium to India. The condition, of course, was adherence to the Nuclear Non-Proliferation Treaty. India, without signing the NPT, was recognised as a "normal" nuclear power by the US. So Delhi has no incentive to sign the NPT for Australia, as it demands uranium equality with China.

During ten years in office, the Howard Government has undergone its own series of conversions on Delhi and Beijing. Those learning curves have been both painful and instructive. The pain with Beijing was really concentrated in 1996. During that first year in power, the Howard Government, almost inadvertently, managed to push just about every wrong button it could in Beijing. Australia was the only country in the region, apart from Singapore, to support the deployment of the US Navy to the Taiwan Straits during the missile crisis. Australia made a symbolic restatement of the ANZUS alliance only a couple of weeks after some significant moves in the Japan-US alliance. The new government sent some positive signals to Taiwan and John Howard cheerfully had a meeting with the Dalai Lama.

Beijing then taught Canberra a valuable lesson – how much pain it could impose. After a few months in office, the Howard Government found itself undergoing the diplomatic death of a thousand cuts. Every single thing that Australia had going through the Chinese system in 1996 ground to a halt. Everything. The Chinese scrapped high-level visits, every Australian doing business in China, whether miner, banker, insurance executive or diplomat, was screaming at Canberra: "Make this pain go away."

So, at the end of 1996, Howard had his Butch Cassidy moment. In the movie, Butch Cassidy is standing on a hill watching the posse that has been pursuing him across the plains of the West. Butch turns to the Sundance Kid and Paul Newman's character says: "If they gave me the money they're spending to stop me robbing them, I'd stop robbing them!" Howard went to his first meeting with China's leader, Jiang Zemin, and delivered a version of Butch's line. The Prime Minister effectively told Jiang: "If you stop imposing the pain on us, we'll stop doing the things which have made you impose that pain on us." And, by and large, the Jiang Zemin–Butch Cassidy pact has held.

When Hong Kong returned to China in 1997, Australia lined up with the Asian states rather than with the US and Britain. Australia, along with the rest of Asia, went to all the ceremonies of the handover. The Europeans and Americans avoided some of them because of the implicit endorsement of the political structure China had designed for Hong Kong. Howard went to Beijing and announced that Australia would no longer take part in the UN human rights process on China. Instead, Australia started a bilateral human-rights dialogue with China. And the next time the Dalai Lama visited Australia, he met no government ministers.

Whenever you hear Howard talk about the Chinese relationship, he will always, somewhere in his default script, put in that line: "We'll concentrate on the things that we can do together and not concentrate on the things which divide us." And, as with so much of the Prime Minister's language, there are a series of meanings and understandings cemented into those words. The "emphasise the positive and sidestep the differences" mantra means Australia has avoided much of the Washington argument about whether China will be a strategic partner or strategic competitor.

The nuances were displayed at the White House in July 2005, when Howard and Bush answered questions on what the rise of China would mean. The US President said he believed that his country and Australia should work together to get China to accept "the same values we share". Australia's leader wasn't interested in joining a values crusade, nor in getting in the middle between America and China: "We don't presume any kind of intermediary role." Australia's relationships with China and the US, he said, were completely separate. Spare us, please, any either–or choices. And the optimistic view put by the Prime Minister is that a "dust-up" between America and China is far from inevitable as a growing China matures and takes its rightful international place.

The differences between Washington and Canberra are often expressed in silence. A key example is the way Australia has quietly sided with China over the biggest economic argument between Beijing and Washington over the past four years. The US has continually harangued China to raise the value of the yuan, saying the artificially low value of the currency gives China an unfair trade advantage and is responsible for much of America's balance of payments crisis. Australia's Treasury, in its 2005 budget strategy, gently dismissed the US position as risible. The Treasury noted that the US current account deficit in 2004 was a record $US666 billion, and offered a series of reasons for this huge imbalance: lack of savings, poor growth in Japan and Europe, and under-investment in some East Asian countries.

Just in case that repudiation of America's China-bashing was too subtle, the Treasury then gave explicit backing to Beijing's position: "The strength of the Chinese economy has led to external calls for greater flexibility in its exchange rate, particularly from the United States. However, a more flexible Chinese exchange rate is likely to have only a limited impact on global imbalances. Broader liberalisation of the capital account should be approached cautiously and coincide with a further strengthening of the Chinese financial system."

No wonder some of the sharper-eyed policy wonks in Washington have started to worry about how far Australia is straying into the Chinese sphere in areas apart from the alliance.

The learning curve Australia has followed with China over the past decade is slowly being replicated with India. Beyond the bromides about the links that run from parliamentary systems to playing fields, the India–Australia diplomatic conversation has been strangely sterile. Australia has been slow to recognise the essential Indian dimension to its Asian strategies. New Delhi's standard view has been of Australia as an obedient member of the Western alliance, happy to act as a cipher for the US. The clash of temperaments and world views reached its low point at the United Nations in 1960, when Nehru took the rostrum to savage an earlier speech by Robert Menzies, arguing that Australia's views on colonialism and the Cold War could not be taken seriously. Menzies wrote furiously to his wife that Nehru was poisonous, sneering and grossly offensive: "All the primitive came out of him."

Four decades later, Australia still had trouble placing India. The Howard Government's first foreign affairs white paper, in 1997, didn't rank India as one of the states that "most substantially engage Australia". And only six months after India announced itself as a nuclear-weapons power with five bomb tests in 1998, Australia's Foreign Affairs Department was able to scrap its separate South Asia and Indian Ocean branch as a budget measure. The contrast between the concentrated focus Australia has given China with the somewhat dilatory nature of the engagement with India is one element in New Delhi's frustration with Canberra. The purpose of Howard's visit to India this March was partly to bury some history as well as lighting the path of future history with one of Asia's pivotal states.

The change in India's significance can be told through two multilateral moments. In 1997, Australia helped ensure that the Asia Pacific Economic Co-operation (APEC) door was closed on India for a decade. By 2005, Australia was slipping into the first East Asia Summit using the diplomatic opening created by India.

At the Vancouver summit of the APEC forum in 1997, India was one of eleven counties seeking to join APEC. Only three got in: Russia (sponsored by the US), Vietnam (courtesy of ASEAN) and Peru (backed by Mexico and Chile). The snub for India was that a ten-year moratorium was then imposed on new APEC members. Closing APEC for a decade was a decision that came out of the daylong leaders' retreat. Howard said he would "very strongly" support the ten-year freeze. The Prime Minister was modest about his own role in ensuring that the moratorium was a decision announced from the summit: "Well, it came up earlier in the discussions and one of the leaders reminded the meeting at the end that it [the freeze] should be included." India's view was that the encouragement it got from some APEC members like Singapore, Malaysia and South Korea had been vetoed by a jaundiced Australia.

By 2005, it was impossible to conceive of the Asia Pacific Climate Partnership or the East Asia Summit without India's involvement. In drawing up the guest list for the first Asian leaders' meeting in Kuala Lumpur in December 2005, China took a narrow view of who should be at the table. Beijing said it preferred the ASEAN-plus-three formula: the three powers from North-East Asia – China, Japan and South Korea – with the ten South-East Asian members of ASEAN. ASEAN could not embrace that formula because it would have excluded India and given China a form of veto over a process supposedly run by ASEAN. Admitting India – a geopolitical must – meant it was easier to invite other states beyond East Asia: Australia and New Zealand.

India, with a middle class that now outnumbers the population of the US, doesn't have to worry anymore about getting invitations. India matters as much as China in everything from clinching a deal at the World Trade Organisation to launching the next stage of the Kyoto process in Montreal.

Indeed, as many of the developed states confront the reality that they can't achieve their Kyoto targets over by 2012, the developing economies on the threshold of Kyotoland have even more room to manoeuvre. The Conservative government that took power in Canada in January 2006 is already showing wobbles about the pain that will be imposed by Kyoto.

Standing outside Kyotoland, the problem for Australia is finding a true role for a non-binding, voluntary partnership. The portents are not good. In that other Asia-Pacific partnership, APEC, the voluntary approach failed spectacularly. Australia championed a fast-track idea in which APEC members would rush to free-trade purity. It was called "early voluntary sectoral liberalisation". That vision died at the APEC summit in 1998 when Japan refused to offer up any voluntary liberalisation in agriculture. Australia's Trade Minister at the time, Tim Fischer, commented that the issue could have destroyed APEC if the argument with Japan had been pushed over the brink.

In trade, Australia has discovered over many decades that the toughest issues can be tackled only when everything is negotiated together in the multilateral system operated by the World Trade Organisation, with binding, legally enforceable rules. In the world of climate change, apparently, different rules can be made to work. Kyotoland is beset by mirages as well as hot air. ■

Graeme Dobell is the foreign affairs/defence corr領ent in Canberra for Radio Australia and ABC Radio. He is the author of *Australia finds home : the choices and chances of an Asia Pacific journey* (ABC Books, 2000). His essay 'Diplomatic compliance' was published in *Griffith REVIEW 1: Insecurity in the New World Order*, Spring 2003.

Policy: **Precautions for the day after tomorrow**

Author: **Michael Heazle**

Climate change is both a major policy issue and a part of popular culture. For many, it has become synonymous with "global warming", catastrophic weather events, submerged countries and a general end to life as we know it today. This view reflects a mainstream acceptance – in the developed world at least – that the Earth is warming: a trend caused by increasing carbon dioxide and other greenhouse-gas (GHG) levels in the atmosphere, with serious consequences for future generations and a sense that, because it is our fault, we can and must do something to avoid, or at least mitigate, the possibly catastrophic consequences.

But uncertainties over the nature, causes and potential impacts of climate change are plentiful and cannot be dismissed in any serious discussion of what global climate change may or may not represent. Supporters of the mainstream global-warming view are able to cite climate-change "facts" that are largely (though not entirely) uncontested, such as increased levels of carbon dioxide in the atmosphere and a general warming trend in global temperatures over the past 150 years. What is far less clear, however, is what these "facts" actually mean and the extent to which they are significant in relation to our existence and the natural environment.

And therein lies the rub. How confident are we, for example, that the "global-warming" scenario is an accurate representation of what is happening today and what will happen tomorrow? What is our confidence for or against global-warming predictions based upon? And, most importantly, how does our confidence in such predictions stack up against the costs of taking precautions today against the predicted (but unknown) future costs of global warming?

The policy problem, as framed by the current global-warming debate, is essentially a choice between accepting substantial economic and political pain today by making emission reductions (high level of certainty) in the hope that doing so will significantly reduce the possible costs of global warming in the future (comparatively low level of certainty), or adopting a mostly business-as-usual approach in the hope that the majority of scientists have got it wrong (for example, the recent Asia Pacific Partnership initiative).

The stakes are high, as are the associated uncertainties. But, because the stakes are so high, it is imperative that we do not make the error of confusing "consensus" with "certainty" or "scepticism" with politically motivated "contrarianism" in the course of debating the appropriate response. Neither scientists nor policy-makers are able to view the potential outcomes and the impacts they may have in a politically neutral or unbiased way; everyone has their own value-based assumptions about what is good, bad or more, or less, important and these assumptions cause us to interpret uncertainty and risk in different ways.

The real policy challenge, then, is to recognise these inherent biases and balance them as best we can. And the best way of doing so is to adopt what Bertrand Russell called a "middle position" on scepticism. In his book *Sceptical Essays* (1928), Russell advised: "Even when the experts all agree, they may well be mistaken. Einstein's view as to the magnitude of the deflection of light by gravitation would have been rejected by all experts twenty years ago, yet it proved to be right. Nevertheless the opinion of experts, when it is unanimous, must be accepted by non-experts as more likely to be right than the opposite opinion."

The problem is that expert opinion is very seldom unanimous; in fact, the more important the issue and the greater the political costs involved, the less likely unanimity becomes. The best that can be hoped for is a majority consensus, which is a good deal less comforting than unanimous opinion when important decisions need to be made. In this situation, Russell advised "that when they [the experts] are not agreed, no opinion can be regarded as certain by a non-expert".

So, as a climate-change "non-expert", and at the risk of being labelled a "contrarian" or "heretic", I will take a sceptical approach to the global-warming debate and initiatives, not to stonewall or delay measures against potential climate-change dangers, but as a way of moving beyond the current political quagmire and developing alternative approaches.

The existing debate is unproductive because it is focused on the unresolvable question – for the time being at least – of "who has got it right?" rather than the more relevant question of "how do we best cope with getting it wrong?" The solutions this debate has offered up so far are either impractical or ineffective (or both), since they are generated mostly by unrealistic expectations and political posturing – as illustrated by the most vocal supporters of the congenitally flawed 1997 Kyoto Protocol, on the one hand, and an unwavering resistance to any climate-change initiative that may interfere with the economic and political interests of some governments, in particular the United States and Australia, on the other.

A far more appropriate response to climate change, and the threats it may or may not involve, is to hedge our bets by concentrating our efforts on replacing fossil fuels with a more diverse and cleaner array of renewable-energy sources as quickly as possible.

Doing so would not only greatly reduce human carbon dioxide emissions but also provide, with a relatively high degree of certainty, a host of additional benefits that could be realised regardless of all but the most catastrophic climate-change consequences and with far less economic and political damage. Prioritising the development of renewable-energy sources would enhance the ability to manage and adapt to future climate changes and still pay dividends, even if global-warming threats fail to occur, or occur despite carbon-emission reductions.

Reducing carbon emissions is not a bad idea with or without the threat of global warming; they do contribute significantly to air pollution. But the Kyoto Protocol, as it stands, is largely symbolic and does little more than state the view that people need to reduce carbon emissions. It is flawed by its narrow emphasis on reducing carbon dioxide levels and its failure to address the broader uncertainties and potential costs involved.

The extent to which climate-change-related problems are, or will be, primarily the result of carbon dioxide increases caused by people is still very unclear. Issues of natural climate variation and the impact of aerosols, water vapour, clouds and sun spots, in addition to the causes and role of naturally generated methane, to name but a few areas of concern, have all raised questions for which scientists and their climate models remain unable to fully account.

The question, then, should not be all about who has got it right; we should also be thinking about how we can develop a strategy that best manages the risks involved with getting the causes and potential effects of climate change wrong, at least until we are in a position to more confidently discuss what is or isn't going to happen.

Research recently published in *Nature* (January 12, 2006) further illustrates the likelihood of *knowledge* suddenly morphing into questionable assumptions. Four European scientists have concluded that, contrary to conventional scientific wisdom, large amounts of methane (an important GHG) are produced by living – instead of decaying – terrestrial vegetation. As all good research should, this study both questions what we think we already know and raises new questions that we haven't previously considered. One of the questions it raises in the context of global warming is the usefulness of mitigating carbon emissions by using forests and reforestation projects as "carbon sinks", one of the Kyoto Protocol's major initiatives, if forests are contributing rather than only absorbing GHGs.

Another emerging problem is that, even if we accept the current consensus on the causes and possible impacts of global warming, there is little agreement over the kinds of carbon-emission reductions needed to create significant relief. Supporters of the so-called mitigation approach, which focuses on emission reductions and is central to the Kyoto Protocol strategy, are now being undermined by warming and GHG retention-rate estimates and impact assessments, indicating that nothing short of major short-term reductions will be effective.

One such study appeared in a report on the *New Scientist* website (NewScientist.com) on February 3, 2005. The report concluded that the European Union's target of limiting global warming to two degrees "now appears wildly optimistic". The study claims that if GHGs are to reach "safe levels", current global emissions need to fall to between thirty and fifty per cent of 1990 levels by 2050 (the Kyoto Protocol aims to reduce the collective GHG emissions of industrialised countries by 2012 by just over five per cent compared with 1990 levels).

If such studies are reliable, the economic and political costs of trying to avoid the worst of global warming are looking more and more likely to be so great as to be unacceptable to all but the most devout environmentalists; most people quite simply could not or would not tolerate the serious economic effects that major short-term emission reductions would involve.

Given the many significant uncertainties that characterise our understanding of the global climate and the effects of our interaction with it, it is absurd to argue that people, especially in developing countries, should accept major economic cost and hardship in order to limit one of the many variables that may be behind climate change. Moreover, it is highly probable that adopting such a response would effectively handicap the ability of societies and governments to take adaptive measures against any climate-change impacts that may actually occur.

Proponents of the global-warming scenario consistently advocate a precautionary approach to the potential threat of rising ocean levels, a stalled Gulf Stream, increased desertification and a host of extreme weather events. It is an approach best illustrated by the vexed ambitions of the Kyoto Protocol and its goal of lowering carbon dioxide emissions in the hope of lessening expected impacts of global warming. The longer-term risks posed by global warming, precautionary advocates argue, are so great they negate the uncertainties or short-term costs.

In other words, an ounce of prevention is worth a pound of cure. But one of the problems with the so-called "precautionary principle" approach to future threats and risks – in addition to the challenge of clearly identifying the potential threats and understanding their nature and likelihood – is the fact that attempts to address one set of risks invariably create other, often unanticipated, risks. As one "pro-global warming" scientist recently put it, although his intentions for doing so were somewhat different from mine, "uncertainty is inevitable, but risk is certain".

And indeed it is. Drastically reducing carbon dioxide emissions may well reduce the future risk of climate-related disasters and hardship. But what additional risks are incurred by doing so, and are they more or less of a threat than the risks they were originally designed to mitigate? One illustration of the unavoidable nature of risk in the context of global warming is provided by the geopolitical instability and tension that would almost certainly result from attempts to impose short-term emission reductions on sovereign states, at a time when energy-access concerns are already creating what some are calling "a new great game" of strategy in Eastern and Central Asia, Europe and the Middle East.

Developing countries like India, China and Indonesia are highly sensitive to the economic costs, especially in relation to energy supply, that dramatic short-term emission reductions would bring. As the fall of the Soeharto regime in Indonesia in the late 1990s demonstrated, stalled economic growth can easily lead to destabilisation of the government and even the state itself, with the potential for internal instability to then spill over and compromise the security (both human and national) of other states or entire regions.

Another "risk" emerges from the emergence of nuclear power as a possible solution to the problem of reducing GHG emissions. Not surprisingly, global-warming scenarios have added a new lustre to the tarnished image of nuclear power. Are the risks of nuclear power and waste less than those of global warming? "Yes" is the rather predictable reply from the nuclear-power lobby, but how big a threat are modern nuclear power stations and who is going to live near them? Equally worrying is the issue of where the waste will be stored, not to mention the question of how much of it will actually become weapons-grade plutonium rather than waste, particularly if fast-breeder reactors become the preferred generators.

If the consequences of global warming are even half as bad as some scientists would have us believe, do the risks posed by nuclear power then become acceptable, and if so, by whose reckoning? A precautionary approach, rather than eliminating or reducing risk, merely changes the kinds of risk we are exposing ourselves to. The question then becomes one of which risks we want to avoid and what we are prepared to sacrifice to do so.

It is a mistake to be focusing on emission reductions, since there are far too many uncertainties and risks involved for us to act with any reasonable degree of confidence in identifying and then selecting those we prefer to face or avoid. Supporters of the mainstream global-warming view often argue that future generations will never forgive us if we fail to act against the future consequences. That is no doubt true, but this argument assumes that we know what the consequences are and also neglects the possibility that drastic action today – on the basis of untestable assumptions about the future – may also have consequences that our great-grandchildren will find equally difficult to forgive – development failures, worsening poverty, neglect of other pressing environmental and social issues, and greater risk of military conflict.

Critics of the Kyoto Protocol are right when they argue that its strategies are undermined by too many questionable assumptions about the likely costs and by divisions, even among those who support the global-warming consensus, over how effective the protocol's reductions would be, even if full international co-operation and implementation were possible. Prominent among such critics, unfortunately, are those who prefer to use the Kyoto Protocol's shortcomings as a political foil for having no climate-change strategy rather than as a reason for developing an effective alternative.

During the inaugural meeting of the six-member Asia Pacific Partnership group (AP6) in January, two of the Kyoto Protocol's biggest critics, US President George W. Bush and Australian Prime Minister John Howard, talked up the importance of developing renewable energy sources as a way of combating global-warming threats without incurring crippling economic penalties. Their actions failed to do their words justice. Howard, for example, made it clear that the

Australian Government remains committed to fossil fuels, calling them "an enduring reality for our lifetime and beyond", and therefore places a comparatively low priority on ensuring their replacement by alternative energy sources any time soon.

Of the $100 million Howard has dedicated to the partnership over the next five years, Australia will contribute only $5 million a year to renewable-energy projects. This, according to a government press release, is in addition to the $200 million the government claims it has already invested in developing renewable energy. (Meanwhile, $500 million has been "invested" in so-called "low-emission technologies".) The US government, which spends more than $US400 billion on its military each year, committed a meagre $US52 million from its 2007 budget, subject to approval by Congress.

Both governments essentially used the AP6 as cover for dodging global warming by announcing their intention to pass the job of developing and implementing new energy technology to the private sector. Like the Kyoto Protocol, AP6 is looking more like an exercise in symbolism, the partnership as a disguise for effectively doing nothing.

Talks on the framework to replace the Kyoto Protocol when it expires in 2012 began last year, and the early indications are that the blueprint is in trouble and may be shelved. Among the possible alternatives is the idea of replacing the existing focus on crude emission cuts with targets for the development and implementation of renewable-energy targets, an encouraging sign that at least some governments are beginning to treat renewable-energy technology as a serious alternative for dealing with climate change.

Climate-change research, in the meantime, should continue with greater support from governments in the hope that future results will better account for the many uncertainties we now face and allow more informed policy decisions in future. The Kyoto Protocol should be scrapped and replaced by an international treaty that reflects a real and determined commitment to develop alternative energy aimed at replacing fossil fuels in the short to medium term. By doing so, we will not only hedge our bets against the potential effects of climate change and our responses to it, but also build an enduring legacy for future generations that will "certainly" be appreciated. ■

References are at www.griffith.edu.au/griffithreview

Michael Heazle is an Australian Research Council post-doctoral fellow with the Griffith Asia Institute. His book *Scientific Uncertainty and the Politics of Whaling* will be published by the University of Washington Press in July 2006.

Reportage:
The brown peril
Reporter:
Chip Rolley

Image: Kaye Kessing / The Terrible Truth / part of the "Battle for the Spinifex" series. / Courtesy of the artist

The brown peril

As a student of Mandarin at Beijing's Tsinghua University in 1999, I made almost monthly visits to the home of a retired academic, who would arrange the purchase of books for my university's library. Visits with Mrs Liang and her husband were always occasioned by pangs of fear; fear that I might mangle my Mandarin or find myself sitting in awkward silence, unaware they were awaiting a reply to a question they'd asked. These visits were lessons for me not only in modern standard Mandarin but in modern Chinese etiquette.

My health and weight would be scrutinised by the pleasantly plump Liang and a seafood dinner would be promised. "You have wonderful seafood in Australia," she allowed. "But you don't know how to cook it. You wash it and wash it and wash it until there is no more flavour." Meanwhile, bowls of individually wrapped snacks would be pushed my way while I'd freeze under the arctic blast of their Fujitsu wall unit and slide off the plastic-covered leather chairs. Liang's home was, by Western standards, a humble flat. But it was very comfortable and well appointed and Liang was clearly proud of all it represented about herself, her family and what was available in China today. She was an ambassador extraordinaire for the *gaige kaifang* (reform and opening up) period, launched by Deng Xiaoping in the wake of the Cultural Revolution.

"What do you think of my home?"

"Lovely." (My Chinese friends described anything remotely good as "lovely".)

"Is it too warm? We can adjust the air-conditioner."

"How convenient," I'd marvel, staring at the remote. (If something's not "lovely", it's "convenient", perhaps the ultimate compliment in a society too long accustomed to the opposite.)

Every time I visited, she made sure I noticed her hardwood floors. "The timber is from Indonesia," she'd attest, stamping the floor with her stockinged foot.

The aspiration of middle-class urban Chinese is not so much a four wheel drive as it is a hardwood floor.

Fast forward four years and I am teaching English in Shanghai. It's hairy crab season. Mr Wu, my landlord, stands at the door of my flat with a plastic bag full of crabs, some struggling for freedom from the hemp twine that binds them. "These males are intense! Real fighters!" he laughs, rushing them into the kitchen area. But it's the two females that will offer up the delicious roe that is the ambrosia of the hairy crab. After Mr Wu takes command of the wok and boils the fighters alive, we settle down to eat. A sweet soy and rice-vinegar sauce completes the ultimate Shanghai culinary experience. The best hairy crabs from Yangcheng Lake are shipped overseas and to Hong Kong, leaving locals scrambling for counterfeit Yangcheng crabs. (DVDs and handbags aren't the only fakes sold in China and the local fisheries department had to devise an authentication tag for genuine Yangcheng crabs. The tag itself was later said to have been immediately counterfeited.) Mr Wu is not even pretending to be offering me Yangcheng crabs. "These are farmed," he says.

"When I was a child, we used to catch everything straight from the Huangpu and eat it that night," he says, as he demonstrates the finer points of sucking the juicy meat out of the creature. It shouldn't be this easy to find someone nostalgic about the 1960s, when the Cultural Revolution ripped apart China's social, economic and cultural fabric. But here's my landlord (my *landlord*) glowing about life's simple pleasures when he was a boy and everyone was equal. He talks about how people used to leave their doors unlocked. "There was nothing to steal!" he laughs. And he rattles off a list of species he and his father would catch in the river that snakes up around the heart of Shanghai, past the Bund and out to the East China Sea. "You couldn't fish there now. If you did, you'd get very, very sick."

Mr Wu did not always bring crab, but he would often settle in for a chat when he made his bimonthly visit to collect the rent for the modest flat I took up while I was teaching in local city high schools. It was a shabby building far enough from the heart of the old French Concession (the hotbed of expat Shanghai) and smack in the middle of urban Shanghai daily life. Real Shanghai, I'd tell myself.

Unlike the manicured tree-lined streets of expatland, or the shopping arcades of boulevards to the north, there was life on my streets. Real life. Every manifestation of it. My window on the thirty-first floor gave me the industrial view of the gantry cranes at the Jiangnan Shipyard, Shanghai's hub industrial port. The new extension emerged above People's Hospital Number Nine as hammer clanked against metal and sparks sprayed twenty-four hours a day in a furious rush to meet the deadline.

My route to one of the schools led me through the morning ablutions and breakfast chats of those of my neighbours who could not afford a high-rise flat. So many seemed to live on the street, especially in the blistering, withering summer months, when *de rigueur* street wear is pyjamas (from thin diplomat-style to patterns of Winnie the Pooh, Hello Kitty and every flower you can imagine). This one sings out the hack-and-spit morning call of China while that one tucks into a fried bread stick dipped in soy milk and another squats over the gutter, bowl in hand, brushing his teeth.

When I'd assign my students the environment as a "conversation" topic, they'd drone well-rehearsed monologues about how clean the air has become in Shanghai. I'd glance out the classroom window and on some days could not see across the street for the smog.

On another rent-collection visit, a stinking hot day in spring, Mr Wu grabs my shoulder. "You know, we never had these temperatures when I was a boy, 35, 37, 38 degrees. Shanghai never had this kind of heat, even in the summer. Some summer nights you'd have to get a blanket!"

Life has changed for Mr Wu over the past twenty years. China has experienced two decades of extraordinary economic growth and cities such as Shanghai have reaped the benefits – better roads, green-belt parks, better plumbing and shops and boutiques lining its tree-lined boulevards, offering all the capitalist consumerist world has to offer. But even in Shanghai which, according to detractors, has attained its privileged position on the back of poor peasants in the countryside, change has not always been an improvement.

In real Shanghai, I needed the tips I garnered from local teachers at the schools about how to shop in the wet markets and on the street. It's a culture that survives on rumours and secrets about scandals that are never reported, or when they are, too late to be helpful. Both river fish and sea fish are risky, I learned. Some might say the river more so than the sea, given the sea's infinitely complex ability to dissipate contamination and cleanse itself. But precisely because the rivers have become so polluted – widely rated as the most toxic in the world – river fish are farmed in a cleaner environment than the sea fish. Or so we're told.

"Always watch the local women in the wet markets. They'll be the first to know what's wrong with what fish. Shop where they shop."

"Be careful of the rice you buy. There have been reports that some is soaked in sump oil to make it shine."

"And never buy from street stalls." Just a few months before I arrived, there were reports that some street vendors selling "smelly tofu" were making the fermented bean curd out of plaster and white paint.

Living in real Shanghai was a lot of work. But a lot less than in other parts of China. The environmental cost of China's development is felt across the country. Sometimes softly, in a nostalgic lament; at others, violently, as guerrilla grannies take to the streets.

It's a spring day in Hangzhou in nearby Zhejiang province, and a young local internet writer joins me on an island-hopping journey on the famously scenic West Lake. We've met through friends and our sightseeing is an opportunity for him to tell me some of what's been going on – and about which papers can be relied upon and which cannot. Peasant uprisings and rural discontent are foremost in his mind. He tells me the story of nearby Huankantou village, which had recently fallen under a media blackout after villagers rioted in protest about a local chemical factory. Some said it was built on contested land and others said it was the cause of deformities in newborn children. A brigade of local grannies were later reported in the international press to be showing visiting journalists their trophies from the conflict – battered police helmets and riot shields they'd seized from the police, who scurried away.

How does one give a sense of the discontent, the unrest in China today? According to official statistics reported by China's Public Security Bureau, in 2005 alone there were 87,000 "mass incidents" (to employ the government's Orwellian parlance). There were 74,000 the year before.

A number of these incidents were centred on grievances about inadequate compensation (sometimes due to embezzlement) for homes and livelihoods destroyed by large-scale development projects. Others, as in Huankantou, focused on aspects of the environmental contamination that has become a feature of China's development.

According to *The Wall Street Journal*, in Qingzhen in the southern province of Guizhou, contamination from a coal-run power station and a chemical factory has polluted the water system, not only causing visitors to gag at the smell, but possibly leading to a number of cases of nervous shaking fits and stomach cancer. The local rice reportedly turns the water it is washed in black and tastes sour after it is cooked.

Tales of environmental catastrophe – both those in the official, controlled media and those that emerge in hushed conspiratorial conversations – are now a staple of the Chinese economic development story. Debate about China's pace and style of economic growth percolates on the internet's multiplying weblogs, some blocked, some not. Zan Aizong is an editor at the newspaper *China Ocean News*, and also writes frequently, and independently, on the internet. In one piece, Zan gives simple, and apparently all too common, examples of environmental crisis in his own province, Zhejiang, which boasts China's most successful market economy, enjoying ten straight years of extraordinary economic growth. People who drive on the 104 State Highway past Xinchang county in Shaoxing municipality all complain of a noxious smell that forces them to roll up their windows. The villagers nearby say that at night as they go to sleep, they don't dare open their windows. When Zan stayed there, his hotel sat in the middle of a string of pharmaceutical factories. The surface of the Xinchang River and its surrounding creeks emitted a putrid smell. A nearby village's residents were told the well water was not potable. "You cannot drink the water, so naturally you cannot water livestock or irrigate crops," writes Zan. "Who takes legal responsibility for this?" (Zan's piece was downloaded in Australia. It is uncertain whether it is accessible in China.)

Critiques of China's dirty development seep into everyday conversations. One friend tells me he thanks his lucky stars every day that he lives in Beijing and not in one of China's interior cities. When I tell him about friends who said that on a recent trip to Xi'an the air pollution was so black and thick they stayed in their hotel rather than make their way out to see the famed terra-cotta warriors, his eyes light up. "Xi'an is cursed," he says. Even among its educated elite, there exists in China reliance on superstition in a way that startles. I look at him sceptically, and he does indeed show some embarrass-ment. But then he proceeds to tell me the story of Qin Shihuang, a familiar enough figure to anyone who has seen the latest crouching-tiger-house-of-hidden-hero flick. In addition to having hordes of clay warriors placed in nearby tombs to defend him in the afterlife, according to China's earliest extant historical account by Sima Qian, the first emperor to unify China has a replica of China in his as yet unearthed mausoleum – a necropolis over which he could rule. Rivers in the necropolis flow with mercury instead of water. It's an accurate enough forecast of the real twenty-first-century China, where at least one provincial study indicates mercury levels in local fish are eighteen times what is considered safe by the Chinese government. "Xi'an will always have a pollution problem because it is cursed," my friend says with renewed authority. In fact, mercury does show up in soil samples near the tomb at a higher level than in other nearby areas.

Xi'an's curse may be something more systematic and persistent than the current localised mercury contamination around Qin Shihuang's mausoleum. Today, according to a multinational scientific study reported in *The Wall Street Journal*, mercury contamination emanating from China's coal-fuelled power stations is carried around the globe by atmospheric currents and appears in samples taken in the United States. It rains down, contaminates wetlands and river systems, and seeps into the food chain. Some scientists in the US claim more than thirty per cent of mercury contamination there comes from China and other countries.

Perhaps nothing has dramatised China's environmental crisis more than the eighty-kilometre toxic slick of benzene that made its way, in late 2005, down the Songhua River, through the north-eastern Chinese city of Harbin and into Russia. Harbin (if known to Westerners at all) hitherto had a secure reputation as a popular winter tourist destination, marvelled at for the fiery brand of *baijiu* (grain alcohol) that comes in a bottle the shape of a hand grenade and its annual ice-lantern festival, where a park near the Songhua fills with ice-sculptured palaces, castles, slides and labyrinths, lit from within, behind and beneath in lurid, hallucinogenic colours. Not even China's notorious coal-mining accidents (in 2005 there were on average more than eight a day, killing almost twice that many people) have made as large an impact on world consciousness. It was as if the dirty secret of China's rampant development had finally oozed, incontrovertibly, to the surface. Experts say the fallout from the Songhua's benzene spill, which has seeped into the riverbed, will plague the area for years to come.

One cannot help but think of the catastrophes in other countries that galvanised governments to act, such as the one at Love Canal in the United States. But what, one wonders, will galvanise the Chinese government? Looking at the province of Zhejiang alone, Zan Aizong says the cases of environmental pollution by large enterprises are "too many to count". China has too many Love Canals for any one to stand out from the rest.

However much an international media event Harbin turned out to be, for me it is the cumulative effect of the events "too many to count", more than any single catastrophe, that has the greatest impact. Anecdotes of the human toll suffocate – the shakes, the itchy eyes, the parched livestock, the Amityville rice which turned the water black, the father and son who collect water samples and hunt for polluting waste outlets in the Huai River in Central China. Statistics smother, too. Sixteen of the world's twenty most polluted cities are in China. Seventy per cent of China's rivers are too contaminated to drink from.

The economic impact of China's development is nothing short of revolutionary. For some analysts, economic development in the US at the turn of the last century provides the best precedent in terms of the influence China's changes will have on the world. The same might be said for the social impact. Chinese officials have been known to point to the US when the dirty style of its development is questioned. (After all, it's the Americans they are emulating in their consumerist, high-demand, fast-growth economy.)

Reading about the odour wafting up from the Xinchang River, I'm reminded of my early teens when I read *The Jungle*, Upton Sinclair's incendiary social-realist novel that exposed the Chicago meat-packing industry and inspired a raft of labour and health and safety reforms (and ensured the average American at the time would never look at sausage the same way again). The stench of Sinclair's American industrial revolution wafted up from the very bitumen of the streets. China is in its jungle now.

Australians can't afford to look too closely at this brown peril engulfing China; such is the extent to which China's frenzied economic development ensures our wellbeing. The Australian share market continues its record-breaking run largely because of surging trade with China: the export prices Australia receives for its coal, iron ore and copper are at three-decade highs and outstripping the prices of imports. Awestruck reports of economic growth stand out in our reading of China. As I write, China just surpassed Great Britain as the fourth largest economy in the world. Between 1980 and 2000, China's gross domestic product quadrupled. Under its National Comprehensive Energy Strategy and Policy Report produced by the Development Research Centre of the State Council, China plans to accomplish this feat again by 2020. That growth promises continuing demand for energy and resources, the lifeblood of "quarry Australis", where the minerals sector alone contributes to about eight per cent of Australia's gross domestic product.

The service economies of Europe, the US and Australia cannot come near to providing the growth in resource consumption that an industrialising and urbanising country like China achieves. In 2006, the growth in China's energy consumption alone will surpass Germany's total energy consumption in 2005. And the good news for Australia is that the overwhelming majority of China's electricity is generated from coal – ranging from sixty-six to seventy-five per cent. China builds a new 1,000-megawatt coal-fired power plant every week, its consumption of coal surpassing two billion tonnes in 2004 – a third of the world's total.

Such statistics roll off the well-lubricated tongues of expats packing into the watering holes in Shanghai's old French Concession. Pubs like the Blarney Stone are a retreat from a city that is itself partly a retreat from the Chineseness of China. They're packed with a burly clientele of hotel managers, entrepreneurs and executives stationed in boomtown China. They gripe about the setbacks but mostly marvel at the unbridled opportunity they find themselves presented with.

BHP Billiton's CEO Chip Goodyear might well be their poster boy. In a speech to the Merrill Lynch Global Metals, Mining & Steel Conference in The Netherlands in May last year, he extolled the opportunity: "When you drive down the freeways in China, in Shanghai or Beijing, and you look at the apartment buildings, and outside every window you see an air-conditioner, that's great, because that consumes steel and aluminium and copper. You only buy an air-conditioner every five or six years, but you turn it on every day."

Such are the delights for the exporter of raw materials when the world's largest nation, so long asleep, awakens with Napoleonic force and moment to its longest period of sustained growth. And when BHP smiles, Australia laughs. Practically every way you look at it, Australia wins with China – supplying the raw materials necessary to manufacture the appliances and goods that are the aspiration of China's middle class, and supplying the resources for the power stations that will keep them all running.

The jungle metaphor implies that this is a phase that will some day end (as it did in the US). Once the long stage of converting its millennia-old agrarian culture to an urban-based consumerist one concludes, the ancient ideals of the Tao or the Way – equilibrium and harmony with nature – may stand a chance. But far from the slick urban China most of us know, high up in the Tibetan Plateau – the "rooftop of the world" and source of Asia's key river systems, the Ganges, the Mekong and the Yellow River – another outcome is unfolding.

It is legend that on the banks of the Yellow River, Chinese civilisation, the oldest extant civilisation in the world, emerged. Five years ago, Madoi County, which sits at the source of the Yellow River in north-west China's Qinghai province, was in the grip of a rat eradication campaign. Like so many sparrows killed in the Cultural Revolution, the rodents were hunted down and caught in a mass party-directed campaign. Drought brought the rodents that, according to the *People's Daily*, took over grazing pastures from herdsmen and severely damaged an estimated 1.3 million hectares of grassland. Today, activists from Greenpeace, which recently established offices on the mainland, travel to Madoi on "dancing roads", distorted by melting subterranean permafrost, and report on dry wells, cracked riverbeds and barren fields in what was once one of the most fertile areas of China. A once self-sufficient community survives on government handouts. "The roof of the world is melting," says Greenpeace.

China is the world's second-largest emitter of greenhouse gases and is forecast to overtake the world's largest, the United States, in twenty years.

The larger impact on global warming begs the question as to whether Australia's economic wellbeing is also predicated on our own destruction. With real evidence of climate change at our door, there is now very little debate about the link to the build-up of greenhouse gases as the cause. Among the key contributors to the build-up of greenhouse gas are coal and oil-fuelled power stations, petrol-burning automobiles, agriculture and materials-processing industries. Even John Howard qualified his long-standing defiant stance against world opinion on solutions to climate change to host the inaugural meeting of the Asia Pacific Partnership on Clean Development and Climate. The group joins the world's only remaining Kyoto recalcitrants, the US and Australia, with India, Japan, South Korea and China in promotion of global agreements based on "clean technology development and deployment that are effective and comprehensive in addressing climate change". It's the pro-nuclear, pro-coal option not offered at Kyoto, allowing Australia, sceptics argue, to have its coal and sell it, too. Industry assistance meets foreign aid in the Australian Government's plan to inject $100 million into an international fund to help China and India implement clean technologies. Scrubbers that reduce "SOx and NOx" (sulphur dioxide and nitrogen dioxide) emissions – already installed in some of China's coal-burning power plants – could receive wider application there and similar technologies could be introduced. But, unlike Kyoto, this deal does not have any binding emission targets for greenhouse-gas reduction.

For the Chinese government, joining the partnership represents another in a string of commitments to address its greenhouse-gas emissions in the face of its continuing industrialisation and burgeoning demand for energy. It is not the first time it has entered agreements to promote transfer of cleaner technologies. Hand wringing about China's environmental destruction does not only happen in hushed conversations and on controversial websites. The Chinese central government has tried to take the initiative. Few seem to have gone further than the deputy director of China's State Environmental Protection Administration, Pan Yue, who, in an interview with Germany's *Der Spiegel* magazine in March 2005, said: "The environment can no longer keep pace. Acid rain is falling on one third of the Chinese territory; half of the water in our seven largest rivers is completely useless, while a quarter of our citizens do not have access to clean drinking water. One third of the urban population is breathing polluted air, and less than twenty per cent of the trash in cities is treated and processed in an environmentally sustainable manner."

Pan forecast "150 million environmental refugees" due to ecological contamination in his country's western regions and warned of a "major blunder" in the thinking that "a prospering economy automatically goes hand in hand with political stability. If our democracy and our legal system lag behind the overall economic development, various groups in the population won't be able to protect their own interests."

The green movement in China rivals only Christianity in capturing the imagination of the urban intellectual elite. University students conduct (friendly, non-threatening) campaigns against disposable chopsticks. Letters to the editor in state-run newspapers call for greater environmental protection and more parks. The tree planting in Beijing is not just to beautify the city for the Olympics, but to combat the desertification that sees windstorms from the expanding western desert whip through the city every spring.

International non-government organisations such as Greenpeace have taken root alongside a host of local environmental activist groups. For these organisations, co-operation rather than confrontation is the operative word and, for the time being at least, the government tolerates them.

With the passage of its *Renewable Energy Promotion Law* last year, China has a mechanism to provide financial incentives for the development of wind energy and bioenergy. This builds on commitments to renewable energy that have startled environmental activists. China has mandates for renewable energy for rural electrification and targets for how much of its total energy demand will be met by "renewables" by the year 2020: fifteen per cent – something, Greenpeace China notes, the US has not undertaken.

This builds on both its decision in 1994 to develop wind farms as a new power source with regulations to make them commercially viable and its New and Renewable Energy Development Program (1996–2010), a commitment to develop solar, wind, geothermal and biomass energy for power. (China already represents 60 per cent of the world's installed capacity for solar hot water use.) China's goals under the energy development program include a massive increase in the efficiency of energy consumption compared with growth, planning to once again only double energy consumption as it quadruples GDP (as it did from 1980 to 2000), and calls for resource conservation as a basic national policy, giving it the same status as controlling population growth. Finally, a series of legal reforms has increased penalties for polluters, created incentives such as preferential loans and tax incentives for "clean" producers, and the environmental agency has closed down power-plant projects for failure to review their impact on the environment.

Even with the heroic targets in renewable energy and efforts to minimise the damage wrought by the use of coal, the very scale of China's development undertaking means its targets can really only mitigate the damage that would occur without them. According to the energy development program, which sets targets according to three levels of growth, by 2030 coal will still account for most of China's energy needs. In addition, the program calls for a substantial increase in reliance on hydropower.

The way China goes about meeting its hydro targets is cause for concern for some. One of the most tragic of China's "mass incidents" was in Hanyuan County on the Dadu River flood plains just east of the Tibetan Plateau in western Sichuan province. In early November 2004, between 20,000 and 100,000 villagers clashed with police, paramilitary and military units at the Pubugou Hydro-electric Dam project there. A media blackout on the riot makes reliable casualty figures impossible to obtain, allowing wild speculation, with figures as low as seventeen and as high as 10,000 (the latter reported by the news website *The Epoch Times*, which is allegedly funded by Falun Gong). The riots were prompted by outrage that compensation for displacement was too low. Villagers reportedly complained that government officials embezzled the compensation money, leaving villagers with less than half of what they were originally promised. Others said the original compensation plan was not even enough to compensate for the destruction of their livelihood.

"Hanyuan" literally means "the source of the Han", the Han being China's (and, for that matter, the world's) majority ethnic group. Its potential to rise to the level of symbol in a critique of China's style of economic development is seized upon by He Qinglian, an economist and now dissident writer, in an essay in Hong Kong's *Kaifang* (Open) magazine.

The Chinese government is "draining the pond to get all the fish", she says – a proverb that roughly equates to "killing the goose that lays the golden egg". The Pubugou Dam project in Hanyuan is one of hundreds of dam projects throughout China that range in size and capacity from small to the world's largest, the Three Gorges Dam. These projects are not only destroying rivers and local ecosystems, as both local and international environmental activists have protested; He Qinglian argues they are also turning economically self-sufficient farmers into impoverished migrant labourers. Money paid for land seized, even when it does reflect the real-estate market value (and is not embezzled), does nothing to compensate for the destruction of a livelihood. This is, she says, the necessary outcome of China's high-speed, high-energy-consuming economic development.

Educated in the economics department at Shanghai's Fudan University, one of China's most prestigious, He Qinglian established her reputation in 1998 with a spirited and controversial critique of the foundations of China's economic development since the advent of Deng Xiaoping. *Xiandaihua de xianjing* (The Pitfalls of Modernisation) was published on the mainland thanks to the support of Liu Ji, vice-president of the Chinese Academy of Social Sciences and an adviser to Jiang Zemin. While the book was considerably toned down from a stronger critique earlier published in Hong Kong, its message was nevertheless clear: economic reform without political reform is doomed to fail. That it was published at all on the mainland was sign of a loosening of authoritarian controls. (Those controls later tightened again and a subsequent article by He Qinglian in a mainland journal found her demoted, placed under surveillance and banned in the media. However, she continues her critique in overseas Chinese publications, such as *Kaifang*.)

How much of China's economic miracle is based on the impoverishment of its people? For He Qinglian, the number of poor created by this form of economic development is just as extraordinary as the wealth produced. She asks us to consider the result when we multiply the number of people who have been displaced and impoverished by large-scale hydro schemes by the number of such schemes planned.

My students at Tsinghua used to go on and on about Samuel P. Huntington's *The Clash of Civilizations* (Simon & Schuster, 1998). But a new thesis is developing. Or is it a sub-thesis? The Brown Peril is Huntington with teeth. Because of its size, China's energy demand represents a threat not only to its own environment and world climate, but also to the West's continuing demand for energy. The brown peril overwhelming China assumes a larger mythical power that, like the "yellow peril" before it, promises the world inevitable conflict. China is a hungry behemoth roaming the world in pursuit of its prey – mineral and energy resources. Such fears are suppressed to varying degrees in Australia – by players across the political spectrum. We appear to be suspending any concerns we may have about the future geopolitical line-up for the present economic gains.

The Guardian reports that change in China now "gobbles up" global resources and represents "enormous risk" for the world balance of power. Change is happening on a "frightening scale", it warns. US Deputy Secretary of State Robert Zoellick avows there is "a cauldron of anxiety about China". "Analysts" on Fox TV in the US tell us the war in Iraq was indeed to secure oil supplies there – justified by the need to stop China getting them. (With that in mind, *The Australian*'s headline in January 2006 warning "China's oil thirst may plunge world into war" seems a little late.)

Unlike the US, so far China's global adventurism has been relatively peaceful, accomplished by the chequebook rather than the gun. With a trade surplus of $US10.4 billion and foreign exchange reserves of $US745 billion, China can afford to be peaceful.

Its pursuit of oil has led it to strike deals with the axis-of-evil state of Iran and the wannabe Sudan. (All other supply routes are tied up by the US, Japan and Europe.) Sudan owes its oil industry to China, such has been the impact of its investment there. Chinese workers built one pipeline there and China National Petroleum Corporation built and fully operates (again with Chinese workers) Khartoum's major oil refinery. In Brazil, Chinese investment deals include the construction of an oil pipeline for Sinopec, one of China's major state-owned energy companies, a steel mill for the Shanghai-based Baosteel and major infrastructure for the transport of soya, which could contribute to China's need for biofuel. In 2004, China overtook the US as Chile's largest copper buyer. Venezuela has signed a bilateral trade deal with China largely focused on energy exports. Roads, dams and major ports from Angola to Kazakhstan are being built either with Chinese direct investment or low-interest loans tied to deals for Chinese firms.

For some of these countries (such as Brazil and Venezuela), these deals not only offer financial and infrastructure rewards, they are a way to make a play against American hegemony – the developing world's turn to play the China card.

For Australia, the more pressing aspect of the brown peril is whether, and how, Australia abandons Labor's three mines uranium policy and develops supply routes to China with yellowcake. A key component in China's "clean" energy bid will be an increase in energy sourced to nuclear power.

In the short to medium term, it's difficult to see how Australia will not continue to reap the rewards of its now twenty-year courtship of China. Nevertheless, China is exercising its consumer muscle, seeking to set prices, much in the way Japan did in the eighties. Fu Ying, China's ambassador to Australia, talked tough at the Melbourne Mining Club late last year, saying the high coal and iron-ore prices (the latter rose 71.5 per cent in 2004) and a source country's "political environment" will force China to "be careful in where it chooses its source of supply". Is this a savvy consumer simply trying to get the best price? Or does this foreshadow stronger attempts to control supply? Will BHP soon be adding another suffix to its name?

It's difficult to gauge what may come of the pressure rising resource prices place on an industrialising China; difficult to untangle what of the brown peril is real – what is a genuine politico-security concern from what is convenient scaremongering based on irrational (can we anymore say race- or culture-based?) fear. Is the brown peril a myth? Is it merely the hysterical flipside of the enthusiasm of the China growth junkies who populate the financial pages? Geo-strategic nervosa is a peculiarly American disease. How much of the brown peril has to do with the fact that, ten years after extricating itself from its extraordinary debt to Japan, the US is now again in debt to an East Asian country. Fear of China's growth sits uncomfortably with any sense of natural justice. The world is on its climate-change precipice not because of anything China did, but because of what has been done by the West, particularly the United States, where per capita energy consumption is seven times that in China.

In October 1999, China launched a week-long celebration of the fiftieth anniversary of the establishment of the People's Republic. I had been there for a month. September in the city had been oppressive, the lingering summer trapping in the pollution of the factories that surround the capital. And then October 1, National Day, arrived. Autumn is typically the most temperate season in Beijing, but in 1999, literally with the turn of the clock to October 1, the air that had been thick and stifling was crystal clear, the sky that had been harshly grey was bright blue. Something was amiss. Overnight China's development had come to a halt. Just for the week. ■

Chip Rolley, a freelance writer based in Sydney, is writing a book about contemporary Shanghai. The names of some people in this essay have been changed.

Essay:
Sunset ports
on the new
trade routes
Author:
Stephen
Muecke

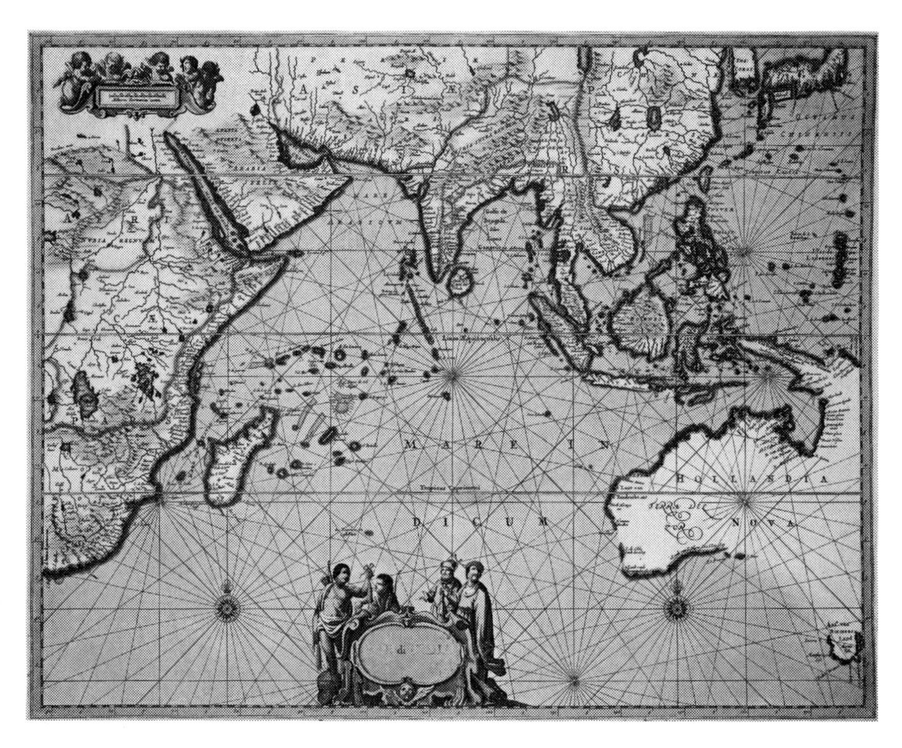

Image: Map by Jansson 'Mar di India [1700]' / Courtesy of the Dixson Library, State Library of NSW.

Sunset ports on the new trade routes

When Ernestine Hill, the pioneering Australian journalist, visited what she called the "ports of sunset" on the West Australian coast in the 1930s, travelling rough with her swag and typewriter, she encountered bits and pieces of Australia's maritime history that have since largely been overlooked. For when it comes to the geopolitics of Australia's oceanic surrounds, the Indian Ocean is lapping at the less-significant back door of the national imaginary, while for the east-coaster, every new day dawns over the Pacific, the sun shining from the direction of our all-important easterly neighbours.

"From 1907 to 1914, Whim Creek," Hill writes of a Pilbara port, "with hundreds of employees, shipped 50,000 tons of oxidised copper ores to London in the Singapore ships and its own fleet of three-masted barques." And further south, at Carnarvon, she walks on the jetty, "a mile and a quarter in length, [where] five or six ships a month call, to carry away the loading of the six-ton wool trucks, and sandalwood from the desert, that is shipped to the East from Fremantle ..." Here she spies "the weirdest vehicle imaginable, a railway trolley with a mast and sail, tangible evidence of a faux pas of long ago. When the Singapore ships are in, this sober little truck heaves up her square of canvas and, to the amazement of onlookers, goes sailing along the jetty, gay as a pearling lugger in the south-easter, to scatter the seagulls and come to her moorings in the railway yards beside the respectable freight engine."

Well before the roads were built to "open up" Western Australia, the ports were there, extending the colony from Fremantle to Wyndham. The trade was local, to be sure, but the main wealth was gained through the export of minerals, as well as pearl shell from Broome, horses for the Indian army from Australind, jarrah sleepers for railways and, more recently, live animals. The volume of this trade is not the point, and it was certainly exceeded by the other states; the point is that Australia has a history of cultural and trading links to the earlier-established colonies on the Indian Ocean. These ports linked Australia to Calcutta, Madras and Singapore in the early days of the colonies, so that we weren't all that isolated. Geoffrey Blainey's famous "tyranny of distance" depended on the cultural myopia that created the anxiety of separation from a far-distant "home" in Europe. This was hardly a problem if your last home was India, or if you were used to going backwards and forwards on one of those Singaporean ships from Western Australia.

The foundations of my argument lie, however, in the pre-colonial Indian Ocean. This thesis is based on global economic history and suggests a geopolitical reorientation: that Australia's place, in terms of proximity and the weight of demography, is more logically south Asian than it is Pacific; India's population is more than a billion, China's is 1.3 billion, South-East Asia and Indonesia are home to another four hundred million people.

In the future, this geopolitical logic may be underpinned by an economic network that is less in thrall to the "free-trade" rhetoric of the northern global corporations and more interested in an "alternative" global economy of this region, a global economy that harks back to the only one worth speaking about before the rise of the European empires. The Indian Ocean was the nexus of Chinese and Indian trade and a generator, as they were wont to say, of "fabulous wealth". Such wealth can be knocked for a six when a tsunami strikes, or when resources are exhausted, suggesting the need for a new economic order that embraces the significance of cultural and natural value. Few of the current discussions link the economy to the environment. As China and India are congratulated for opening up to world trade, their natural resources are treated as global raw material, and as if they are not already tied up in peasant 'eco-nomic' systems.

In the pre-colonial period, India was the centre of an inter-regional world system, in the sense that Emanuel Wallerstein used the term. For the Arab world and East Africa from the west, and South-East Asia and China from the east, India was the intersection of trade, the imaginary of wealth borne on the waves of the Indian Ocean. This imaginary was eventually shared by Spain and Portugal, and realised in the most concrete way by Vasco da Gama's arrival at Calicut in 1498 and his aggressive demand for a trade monopoly there.

The "fabulous wealth" of the East became less legendary and more attainable, however, when Spain pursued this centre of trade, India, via a western route and unexpectedly bumped into the Americas. Because of this, says Enrique Dussel, "The entire medieval paradigm enters into crisis ... and thus inaugurates, slowly but irreversibly, the first world hegemony. This is the only world system that has existed in planetary history, and this is the modern system, European in its centre, capitalist in its economy."

With the precious metal wealth extracted from the Americas, Europe was able to buy its way into the Indian Ocean market and start working towards the "northern" hegemony we know today. Once much poorer than India, and perhaps more driven and less complacent, these European maritime states would cleverly continue to combine military power with trade as they competed with each other and forced new trading relationships, thus inaugurating the great European colonial period and the annexation of lands all over the world. Now that this period is over, and world capitalist hegemony is becoming more evenly networked, the Indian Ocean is re-emerging as a trading bloc. In the wake of the East Asian surge, even a tiny Indian Ocean country, Mauritius, a free-trade zone, can dub itself hopefully as a "little tiger". Indian Ocean Rim Association for Regional Cooperation (IOR-ARC) countries contributed about ten per cent each to world exports and imports over the past decade, economists of the Australia South Asia Research Centre report. The figures for current trade with Australia are minuscule compared with the Indian Ocean of the 1400s when China and India together accounted for more than half of the world's gross national product (GNP). Even as recently as 1820 – in the middle of the colonial period – China accounted for 29 per cent of the global economy and India for another 16 per cent. Today China's share of the global economy is about five per cent and growing rapidly.

Nicholas D. Kristof and Sheryl Wudunn have written of the great Chinese admiral Zheng He (originally a Muslim captured as a boy) in *China Wakes* (Random House, 1995). Between 1405 and 1433, Zheng He led seven expeditions, commanding the largest armada the world would see for the next five centuries. The fleet included 28,000 sailors on 300 ships, the longest of which were 120 metres, majestically ploughing the waves with nine masts carrying red silk (by way of contrast, Columbus 50 years later had three ships, each about 25 metres, and only 90 men). Zheng He travelled to East Africa to trade for ivory, spices, medicines and exotic wood. He was probably not interested in trading with a backward Europe with only beads, wine and wool to offer. Paris probably had a total population of about 100,000 in the fifteenth century, while Guangzhou in China appears to have had twice as many foreign residents: Arabs, Malays, Indians, Africans and Turks, evidence of very early merchant diasporas. In 1986, excavations in Guangzhou unearthed the remains of a large Hindu temple of the Yuan dynasty. The archaeologists who made the discoveries referred to this era as one in which "people from every country lived in harmony … [and] made great contributions to the prosperity, economy and culture of Quanzhou".

Where does Australia stand in relation to this history? We were never part of the old global economy of the Indian Ocean, except for the limited indigenous trade with Indonesians and Chinese as Regina Ganter described in "Turning the map upside down" (*Griffith REVIEW 9*: *Up North*). Australia's trading orientations started off strictly colonial, and only in the latter half of the twentieth century did Australia begin to acknowledge the autonomy and significance of Chinese and Indian markets.

Now, at the beginning of the twenty-first century, these two countries are rivals in the computer and software industries and as the global factory supplying manufactured goods to a world of greedy consumers; they talk up their co-operation in terms of a new "Asian century". When, in April 2005, they settled a long-standing border dispute and signed an agreement designed to double trade over the next five years, the rest of the world began to take notice. As global attention focuses on the importance of China and India and concern about climate change mounts, pursuing an alternative economic model becomes more pressing.

Last year, the Pakistani foreign minister and the Indian commerce minister visited Australia, and the chairman of China's National People's Congress pursued a free-trade agreement. Prime Minister John Howard travelled to India in March 2006 and there have been numerous state and federal trade missions focusing on information technology, minerals (Western Australia), tourism and creative industries (Queensland), and coal, tourism and creative industries (NSW). In 2003–04, Australia–India bilateral trade was $5.86 billion, an increase of more than 50 per cent over the previous year. As journalist Geoffrey Barker has noted: "The trade opportunities in both countries [China and India] mesmerise the Federal Government."

The population of South Asia, led by India, will soon exceed the ageing population of China; India has a huge middle class that is attracting Australian exporters. The Australian diplomatic mind-set will have to change, to accommodate a realisation that Australia can be part of the return of Asian economic power. This could make the colonial period of the seventeenth to the twentieth century look like a temporary interruption in traditional global trade.

In 2001, the Economic Analytical Unit of the Department of Foreign Affairs and Trade released a report, *India: New Economy, Old Economy*. Its analysis shows that, while India was only picking up just over one per cent of Australian merchandise exports at that stage, growth in trade had been increasing by about twelve per cent a year, so that India moved from Australia's twenty-fifth to twelfth largest export destination. The pace has since accelerated.

Pitched in the language of neo-liberal economic reform, this report is addressed to businesses wanting to get in on this market. It is a challenge that has been accepted and is the new focus of corporate expansion. The neo-liberal language is nothing new, being another northern ideological product that both India and Australia have come to share. It has effectively lubricated the gears of Indian business exchange ever since the country floated its currency exchange (devaluing the rupee) in 1991 in response to a serious balance of payments deficit. At that moment, India seriously went global, opening the economy to international trade and finance, and reducing export tariffs and regulations.

This is the context in which India has embraced the global economy. It is a second wave of reform, because the post-independence period saw the subcontinent modernising with a series of Soviet-inspired five-year plans. In the 1950s and 1960s, India sought assistance from foreign investors to gear up its industries for global competitiveness. As India played a supporting role in the global power struggle of the Cold War, its economic capacity was overlooked in the West, which continued to focus expansionist dreams on China. Now that China has become an indispensable part of the global economy, the factory of the world, attention has switched to India. The scale of the domestic market, the capacity to integrate with the global economy and the importance of fostering a (democratic) regional counterweight to China have combined to restore India at the crossroads of global trade. It is being restored to its traditional role.

The difference is that the emerging power of India is occurring at a time when a new paradigm of trade is emerging, one that values the local and seeks a balance with natural resources. The development of South Asia has the potential to pull these strands into the development mix in a way that has not happened before.

An example from the Goa–Kerala areas on the south-west coast of India, of the fishing industry, demonstrates this potential. The compiler of the activist sourcebook on Goan ecology, *Fish Curry and Rice* (Goa Foundation), Claude Alvarez is anti-global through long experience, and forceful in his support of local traditional practices. For instance, with the "mechanisation" of fishing in the regions, of a total population of 1.3 million in the state of Goa, he estimates that 50,000 people make a living from fish harvesting on the hundred kilometre coastline. But he notes that when the West started to "help" in the early 1950s by increasing the fish catch with new technology, the real motive may have been to create a market for the technology.

The West is not the only villain for Alvarez, who identifies an ideology of mechanisation enshrined in the gospels of Indian development policy – the famous Nehruvian five-year plans. But he concludes the whole exercise was disastrous. Even the Indian Central Marine Fisheries Research Institute shows a much greater return on investment for peasant fisheries than for mechanised fishing with trawlers. So Alvarez's strictly local activism means he can't, or won't, factor in macro-economic pressures associated with globalisation and reform. It is hard to trade off the national need to be in the global economy by devaluing the rupee and going for export markets against the need for locals to continue a peasant economy.

This contradiction played itself out in local politics as fisherfolk opposed the introduction of power boats and trawlers with mass rallies and alternative candidates. While many mobilised to save their livelihoods, others got into the trawling business. In terms of the management of coastal cultures, mechanisation was a monoculture mentality: a massive technology harvesting one product in which any fine-tuned interchanges between coast and hinterland were irrelevant. Perhaps the era of the peasant farmer quietly trading rice for fish on the coast is over, as the people by the beach in Goa rent out shacks to the huge numbers of foreign tourists coming each year. But the tourists, too, must eat, and even if the best fish and shrimp find their way to their tables at inflated prices, the local ecology is under pressure. Indeed, fish stocks in the whole of the Indian Ocean are declining to the point of crisis, like they are in the North Sea, where sustainability strategies are now firmly in place.

Sustainability is not something that most official economic reports factor into their calculations of wealth. Partha Dasgupta, professor of economics at St John's College, Cambridge, on the other hand, helps us think about the relations of economies to ecologies grounded in local places, not floating free in the abstracted realm of reform policy as dictated by corporate globalism: "Economists ... have moved steadily away from seeing location as a determinant of human experience. Indeed, economic progress is seen as a release from location's grip on our lives." In this context, the macro-economic story contradicts the local stories about the love of homeland, as well as travellers' tales about the beauty of place. Each story is told about some "good" (value), crafting a narrative of a place that is in some sense algebraic; a set of calculations working towards a positive outcome, that values production and place. Dasgupta wrote in the *New Statesman* in November 2003: "Economic statisticians interpret wealth narrowly. Wealth should include not only manufactured capital (roads and buildings, machinery and equipment, cables and ports) and what is nowadays called human capital (knowledge and skills), but also natural capital (oil and minerals, fisheries, forests and, more broadly, ecosystems). I use the term 'inclusive investment' for this broader definition of wealth and contrast it with the narrower scope of 'recorded investment'."

The logic of natural capital is one of complexity and interconnectedness, and this logic is destroyed by the imposition of the grid pattern of ownership of parcels of land, where the "free services" provided by nature, as in a stream flowing through a number of properties, or the bounty of the sea, do not enter into the calculations of economists. He writes: "Those who destroy mangroves in order to create shrimp farms, or cut down forests in the uplands or watersheds to export timber, are not required to compensate fishermen dependent on the mangroves, or people in the lowlands whose fields and fisheries are protected by the upland forests. Economic development in the guise of growth in per capita GNP of improvement in the Human Development Index can come in tandem with the decline in the wealth of some of society's poorest members. Rural communities in poor countries recognised the local connectedness of nature's services long ago and devised mechanisms to cope with the problems created by it. A pond or woodland is a system of organic and inorganic material, offering multiple services. This feature of ponds and woodlands makes them unsuitable for division into private property."

Dasgupta, by adding another variable ("nature's services") to the formula for an inclusive calculation of wealth, comes up with a result that contradicts the misplaced optimism about the rise of wealth in developing economies. India, for example, is supposed to be growing at a healthy 2.3 per cent of GNP a head – on Dasgupta's figures – but on the inclusive calculation the average Indian is getting poorer at a rate of half a per cent each year. The warming of the Indian Ocean, with its effect on reefs and fish, has produced an immediate reduction in natural capital. This is going beyond the simple reduction in fish stocks; the crisis is pushing the ocean into an unfamiliar eco-system.

Future trade should sensibly see itself as sustainable, passing on to the next generation at least the "wealth" we still cling to. Economists and business leaders henceforth should invite a new partner to the summit talks, to make nature a signatory to the contract, as French philosopher Michel Serres advocates in his book *The Natural Contract* (University of Michigan Press, 1995).

There is a fairly recent post-humanist philosophy where man more modestly assumes a less central philosophical position in the world. European Enlightenment, though, had man looking up to a singular god, and down to lower species and to nature. In the central and mediating position, between god and nature, he embarked on a global adventure, a forward-marching triumphant modernity. But now nature — or rather 'natures' in the plural —are arguing back, putting forceful arguments about the finitude of resources. Suddenly, the value of local places, villages and streams, coasts and islands, comes to the fore. "Are our thoughts, until recently rooted exclusively in their own history, rediscovering geography, essential and exquisite? Could philosophy, once alone in thinking globally, be dreaming no longer?" Serres asks.

I take my cue from him and seek to integrate value as a cultural and economic term. It is worth bearing in mind that both Australian and Indian thinkers – philosophers and economists – can, and have already, thought outside the "box" of the European Enlightenment, and can embrace both the traditional peasant farmer's concerns and the notion of "alternative modernities". In traditional Indian Ocean trade, commodity value is obviously crucial, and it governs the movement of goods.

The trade in ivory is a stunning case study. Ivory was in abundance in Africa in the pre-colonial period. It was not treated as a precious substance. Resistant to termites, it was used for palisades or cattle yards in villages. This meant that traders from India could do good business, distributing ivory in the then global economy centred in the Indian Ocean, where it would be transformed into a luxury item, worth many times its original value. So a natural material, abundant in certain areas, is converted through the addition of cultural value into a precious material. Now a new set of values asserts itself via the argument that is put by elephants, so to speak, through their human advocates. The arguments trumpeted by the suffering elephants are about sustainability and ecological balance in Africa. By integrating the concept of value, cultural values (an appreciation of the beauty of a carved ivory artefact or, conversely, the compassion for elephants criminally slaughtered) need not contradict or compete with the corporation's need for profit. The corporation and its shareholders must be convinced, though, about the unsustainability of a purely exploitative relationship with nature where the value flows only one way and is assumed to be limitless.

Another example, as I try to rejig the system of values in this narrative of a natural contract, is to think again about competition, which is taken to be "second nature" in business. Indeed, let us admit that here, too, in the "culture" of business, there is also "nature", for in a post-humanist perspective it is far from productive to have strict separation between cultures and natures as if only the former can act, while the latter is passive, unadaptive and without its own histories.

In early Indian Ocean trade, the peaceful practice of the Jains and Banians has been noted by the historians, who have also gone on to point out that the Europeans brought another kind of configuration to trade when they started to blockade ports to force monopolies: their ships had cannon. Henceforth, trade competition was a war, and it still is (as in Iraq, with oil, an inert substance, but a major player). "If we move from war to economic relations," says Serres, "nothing notable changes in the argument."

After the economic battles have been waged, with all of their waste and inefficiency, a contract might be signed between the warring parties to divide the remaining spoils. But nature is only just starting to be present as a signatory to this social and economic contract.

The tsunami in the Indian Ocean is a case of the natural at its most forceful and catastrophic. Normally, the ocean is known and "read" by its coastal peoples in its complex, chaotic and rhythmic patterns. The deep knowledge of monsoons, currents and tides has underpinned a living for millennia for the peoples now unceremoniously ousted from the coast by tsunami and tourism developers.

The ocean literally connects us to our north-western neighbours. The connection was made compassionately by many Australians at the time of the tsunami and may help shape future relations as Australia takes its place in a new global trading order. As the international studies commentator and academic Anthony Burke has commented: "On maps the oceans are crisscrossed with borders, but their liquid depths have no concept of them, and nor do the earth and atmosphere with which they form a dynamic system ... The oceans that connect us – especially in times of desperation and tragedy – show us the way forward. Our fates are linked; let's make a better fate, a common fate."

Let us hope that the compassion towards suffering can also moderate and refine the language of politics and commerce, so that these, too, can include a much greater measure of the health of living things and the environment, creating a wealth that can be shared and passed on. ■

Stephen Muecke leads an ARC-funded project, 'Culture and commerce in the Indian Ocean' at the University of Technology, Sydney. His most recent book is *Ancient and Modern: Time, culture and indigenous philosophy* (UNSW Press, 2004).

Essay:
Riding Australia's big dipper
Author:
Geoffrey Blainey

Image: Aaron Bunch / One Mile Jetty, Carnarvon, Western Australia 1995 [picture] / Courtesy National Library of Australia

Riding Australia's big dipper

Australia's history during the past 150 years has often revolved around fossil fuels. They have affected not only daily life but the rise and fall of Australian states. They have shaped turning points in local politics and international relations. Thus, the outbreak of war between Australia and Japan in December 1941 was influenced by fossil fuel; so, too, was that major landmark in politics, the election victory of R.G. Menzies in 1949 – the start of the longest reign in federal politics.

In the early British history of Australia, firewood was the important fuel. The timber cutter was more important than the coalminer. By the 1850s, with Australia entering the age of steam, coal became vital. Gasworks were built in Sydney and Melbourne, foundries and factories appeared here and there; and all these ventures burned lots of coal. The first short-distance railways were built in Sydney and Melbourne in the 1850s and they preferred coal to firewood. At the same time, a few steamships, competing with fleets of sailing ships, plied the long route between England and Australia, and they burned so much coal in each twenty-four hours that they were in danger of running out of it in mid-ocean. Therefore, they combined sail power and coal power on the long stretches between coaling ports. When the winds blew strongly the steam engines were idle.

The mines of the Hunter Valley, along with the coalmines south of Sydney, increasingly exported coal. By the 1890s, many sailing ships waited daily at the mouth of the Hunter to load the coal – it was ideal for producing steam, and carried it to ports in countries as far away as Chile, Peru, China and the Philippines. Those cheap old workhorses, the sailing ships, usually carried the coal, for they charged low rates. In contrast, the multiplying number of steamships burned much of the coal. It was a strange but valuable marriage, wind and fossil fuels. The marriage so important to Newcastle virtually ended around 1914 with the coming of the First World War and the fast decline of the globe's fleet of sailing ships.

In 1900, coal was not one of the country's top export industries: it was dwarfed by wool and gold. That a century hence, coal would be the leading Australian export was inconceivable. But coal was already vital for NSW. After the discovery of gold in 1851, Victoria had caught up to NSW in population and swept far ahead; but the fact that coastal NSW held, so far as was then known, the great black coal resources of the continent proved to be a terrible long-term blow to Victoria. NSW overtook Victoria in population in the early 1890s; Sydney overtook Melbourne a decade later. Part of the promise and strength of NSW was that, in an era when black coal was becoming vital, it was the dominant source.

When one of the crucial economic decisions so far facing Australia had to be made where should the ironworks and steelworks and all the allied industries be established ? there was only one answer: in NSW, for it held the beds of high-grade black coal. By 1914 the first steelworks were busy at Lithgow, alongside a coalfield on the inland side of the Blue Mountains. A year later, far more important steelworks were opened in Newcastle by BHP, then primarily a silver-lead miner at Broken Hill. The new works poured out their clouds of steam and smoke alongside the river at Newcastle and steadily gave rise to a network of industries that used iron and steel, and coal, too, as their raw materials. Nothing did more to make NSW, and not Victoria, the centre of manufacturing in the twentieth century than the possession of fine seams of black coal.

At the time of its birth in 1901 the Commonwealth was almost entirely self-sufficient in fuels. It mined all its own coal and even exported some. It cut its own firewood, which was a vital fuel for cooking, heating and generating steam; the steam boilers of the great Kalgoorlie goldfield depended overwhelmingly on firewood cut in the district by an army of woodsmen and delivered by little railways. Australia also produced its own hay, oats, chaff and grass for the horse teams so vital on the farmlands and for carting goods in the cities. And the wind itself gave power to the deep-sea windjammers and other ships that survived by carrying away to Europe the exports of timber, coal and wheat from a wide circle of ports stretching from Bunbury to Newcastle.

But this charmed era of self-sufficiency was broken by the internal-combustion engine and its craving for petrol. With its distances and sparse population, Australia was especially suited to motor vehicles. By 1930, Australia with barely six million people was one of the most motorised nations in the world. It owned more vehicles than populous, motor-loving Italy. The petrol pump, worked by hand, was almost a national symbol. There was only one hitch. Australia itself produced no oil.

Ironically, Australians and New Zealanders, whose lands were then totally deficient in oilfields, were already prominent in discovering oilfields elsewhere. William K. D'Arcy, a solicitor in Rockhampton, made his fortune from the gold of Mount Morgan – the first Australian public company (the share-market darling) to pay one million pounds in dividends in the space of a year. He used part of this fortune to finance a search for oil in Iran. This led to the discovery of the first oil in the Middle East in 1908, one of the most momentous events of the twentieth century. Later, a New Zealander, Major Frank Holmes, played a vital part in finding the first oil in the Arabian Peninsula in the 1930s. In contrast, for decades, the search for oil on Australian soil was neither strenuous nor successful.

The fifty years 1901-51, with their growing emphasis on oil, transformed Australia from self-sufficiency in fuels to vulnerable dependency. As the motor car and then the tractor penetrated everywhere, as aircraft became common, as deep-sea ships began to burn oil and then diesel fuel, and as many locomotives eventually turned to the same fuels, Australia became a huge importer of oil and petroleum. The nation's balance of payments, and at times its financial solvency, suffered as a result. Fossil fuels became our biggest import.

In the Second World War, with oil so essential, Australia's whole economy was vulnerable. Its whole way of life and work depended heavily on the oil tankers that plied the long sea routes from the Persian Gulf and other oil regions. Australia was forced to ration petrol in 1940.

Japan was just as reliant on imported oil. The extension of the fighting to the Pacific and South-East Asia was strongly influenced by Japan's quest for oil. A heavy importer of oil, it was largely deprived of that oil by economic and financial sanctions imposed in 1941 a few months before Pearl Harbor by the United States, Britain and Australia, too. Japan's answer to the embargo on its oil imports was to launch its astonishing military attack, which in the space of less than three monthscaptured the great oilfields of Dutch Indonesia and British Burma.

More than four years after the war was over, Canberra continued to ration petrol. No Australian motorist, no carrier, no farmer, no taxi driver could buy petrol without also handing over, at the service station, a petrol coupon. If you were lucky enough to buy one of the very first Holden cars in November 1948 you still had the task of scavenging, somehow, enough petrol to drive away for a long holiday.

Queenslanders led the fight against petrol rationing and in July 1949 a Brisbane garage-owner took his case to the High Court. A Queensland politician, Sir Arthur Fadden, leader of the federal Country Party (now the Nationals), who initiated the public campaign to terminate petrol rationing. He was swimming on the Gold Coast in 1949 when he was confronted by the head of an Australian-owned oil importer, Ampol, who demanded to know why Ampol could buy petrol from Poland and France, but was not allowed by the federal government to import it. Fadden took up the case and it was a vital topic in the federal election of 1949. He became the treasurer in the new Menzies government, which abolished petrol rationing on February 8, 1950.

Australia was still a large producer of coal. But the coalmines became less efficient, less reliable. Back in 1910, Newcastle had been such a cheap producer that it sent a procession of sailing ships loaded with coal to other lands; but, in the next thirty years, the coalmines lost their export markets. Industrial relations on the coalfields deteriorated. During the Second World War, when the nation relied heavily on coal for making iron and steel and munitions, as well as for domestic purposes, the war effort was disrupted by coal disputes. Of the millions of working days lost through industrial upheavals during the war, fifty-three per cent of the lost days were in the coalmines. When France was about to fall and Hitler seemed likely to invade Britain, the coalminers of NSW were engrossed in a long strike.

Nearly every Sydney and Melbourne housewife at that time knew the names of the NSW coalmines because they were always in the news: the coal-miners were often on strike or locked out. In the late 1940s, Melbourne and Sydney periodically suffered from a shortage of gas and electricity for cooking, heating and transport. In 1949, Ben Chifley, a Labor prime minister, had to send Australian troops to the NSW coalfields in the hope of producing urgently needed coal.

Victoria, especially, suffered because it lacked its own cheap energy. Much of its coal was imported from NSW. That coal had to run the gauntlet. Strikes were frequent not only in the NSW coalfields but at the waterside where the coal was loaded and in the little colliers that carried the coal to Melbourne.

Victoria's solution was to develop its own massive low-grade coal deposits. Close to the surface in the La Trobe Valley, the deposits of brown coal required an enormous injection of capital. By 1960, a large part of Victoria's electricity was coming from state-owned brown coal. Similarly, much of the gas used in Melbourne whether in houses or factories, came from the new Lurgi process by which brown coal was converted to gas and piped to Melbourne. It was a sign of the zeal with which Victoria harnessed a second-rate resource that two of the nation's major aluminium smelters – the manufacture of aluminium is a huge consumer of electricity – were built on the coast of Victoria, one near Geelong in the 1960s and the other at Portland in the 1980s. If the mining of black coal in NSW had been efficient, and interstate transport had been, too, Victoria might never have developed an aluminium industry.

The great Snowy Mountains scheme was, in part, another attempt to escape from the chaotic condition of NSW coalmining. The plan was that hydro-electricity, which was less vulnerable to poor industrial relations, would supplement power generated from black coal. The water, having passed through the power stations, would also irrigate inland plains.

The Snowy scheme caught the nation's imagination and still does. Bold in its conception, it was the first post-war haven for a multi-ethnic workforce. As it was financed and controlled by the major governments and was also conspicuously close to Canberra, it gained maximum political publicity during the quarter century (1949-74) in which it was constructed. Tasmania was also bold in turning to hydro-electricity. By the end of the First World War, it operated a government station at Waddamana and a private one at Mt Lyell. Tasmania's great era of dam building was to come after the Second World War, but it did not catch the nation's imagination.

The grand Snowy scheme, in its heyday, was a living national monument. But probably its economic importance was exaggerated. Far more important in giving Australia cheaper fuel and energy were the rising imports of cheap oil and scattered projects of the 1960s and 1970s that mostly arose far from Canberra and were sponsored by the private sector. Australia's power industry was transformed less by the Snowy scheme than by the discovery of oil and natural gas in the Moomba and Cooper basins in Central Australia, the massive discoveries in Bass Strait in the 1960s and in the north-west shelf from the 1970s, by the development of the huge coalfields in the Bowen Basin in Queensland, by the revitalising of the older coalfields of NSW and the continuing development of brown coal in eastern Victoria. While the Snowy scheme fired the public imagination, those other discoveries and projects fired the economy.

Another point can be made with confidence. Without the discovery of oil in Bass Strait, Australia in the 1970s would have been battered by the two oil crises and the worldwide explosion in petrol prices. As a result, the standard of living of the average Australian would probably have fallen. Today, Australia's reserves of oil are diminishing and expenditure on imported oil is rising. But, all in all, thanks to the strenuous revival of coalmining and the expansion of output of liquefied natural gas, Australia is still a net exporter of oil and natural gas on an impressive scale.

Black coal has become the number-one Australian export, a position unimaginable as late as 1975 when wool was still king. Indeed, if coal is subdivided into the coking and steaming coals, these separate products fill first and third places in the nation's list of exports, with crude oil coming fourth. Moreover, the great ports of the continent measured in tonnage are no longer Melbourne, Sydney and the old capital-city ports but the harbours, mostly new and tropical, that specialise in exporting coal and iron ore in huge bulk carriers. Of today's great coal ports, Newcastle remains prominent while the other three are on the central Queensland coast, serving a tropical coal region that mined virtually nothing in 1970 and today produces more coal than England.

While the energy industry in Australia since the end of the Second World War has been revolutionised, with profound effects on the economy and the standard of living, not everybody cheers. The opponents of new forms of energy have multiplied. Back in the age of steam, coal as the main fuel had few opponents. Admittedly, coal smoke polluted the sky and NSW coal was especially noted for its dense black smoke. But passengers with few exceptions still preferred a steamship to a sailing ship and a steam train to a horse-drawn coach. Naturally the miners of coal faced physical danger but that was discounted by all except the miners and their families.

When oil became the new king, it, too, created few enemies: at first, the smog created by the internal-combustion engine was not very visible. These two fossil fuels, coal and oil, were generally applauded because, on the whole, they saved so much hard, relentless human labour and sweat. One of the major changes in the past century is that the majority of the workforce no longer carries out, day by day, hard physical work. For this triumph the fossil fuels deserve high praise.

In Australia, the first major campaign directed by the new greens against a form of domestic energy was not against fossil fuels but against hydro-electricity. That campaign initially concentrated on the banning of several new schemes that endangered places of rare natural beauty, Lake Pedder and the Franklin River in Tasmania. Nor was the second major campaign directed against the fossil fuels: it was against uranium.

Australia has huge reserves of high-grade uranium ores but produces no nuclear energy for its power grid. An influential section of public opinion opposes the mining of uranium; an even stronger section opposes the building of nuclear power stations. It is reasonable to suggest that if the home-grown supplies of energy had remained as deficient as they were in 1950 and if the hydro-electric, black and brown coal, oil and natural gas ventures of later decades had never emerged, then Australian governments would have been forced to sponsor major nuclear powerhouses.In the mid 1960s, South Australia, gravely deficient in oil and black coal, went close to designing and building its own nuclear power station on Torrens Island, near Adelaide. Another power station was proposed for Jervis Bay in NSW. The public opposition to uranium was not yet large-scale.

In the 1970s, left-wing hostility to uranium was mounting. In 1979, at the Australian Council of Trade Unions congress, Bob Hawke suffered a major blow after arguing that the three existing uranium mines in Australia should be allowed to continue. The trade union movement disagreed with him. By a clear majority it denounced uranium. After Hawke became prime minister he gained a limited victory. Two uranium mines were allowed to continue, but that was the limit. It is fair to suggest that the effective hostility to the mining and exporting of uranium was based on one clear fact: uranium was not seen as vital to national survival or economic development. It was much easier to oppose uranium when other sources of energy were being developed in abundance. On the other hand, if Australia in 1980 had produced no oil and natural gas and insufficient coal, uranium would have won far more friends than enemies. Sheer economics would have made national opposition to uranium a dubious luxury.

Fossil fuels initially were not a major target for critics. As the smog increased in Sydney and Melbourne in still weather, oil and coal were periodically denounced as a source of pollution and illnesses, but not yet of global warming.

The Whitlam government was the first to examin the question of global climate change, though from a direction that now seems surprising. The government was stirred by the fear expressed at the World Food Conference in November 1974 – and evidence supported that fear – that the world was cooling and that such a change, if true, might damage Australia's vital rural industries. As a result, the Australian Academy of Science was invited to report. Setting up a committee of eleven scientists, led by C.H.B. Priestley, it released its report *On Climatic Change* in March 1976. Scholarly and cautious, it treated global cooling as an unlikely possibility, saw the hazards for Australian farmers of relatively small shifts in climate, and was emphatic that climate in a typical century was more variable than people realised. The idea that the world by the late 1990s might become appreciably warmer was not then envisaged widely in either Australia or the northern hemisphere.

Now global warming rather than cooling is seen as the danger. And yet the Australian opposition to its country's widespread output of fossil fuels, the massive export of coal, and the might of the local metallurgical industries remains relatively muted. The reason is pretty obvious. Australia is one of the great miners in the world. Such industries are vital to our standard of living. But the whole question of Australia's role in minerals requires more than that partly-true economic explanation.

There exists a long-term palliative for global warming – massive diversion of resource to the production of nuclear energy. If there was such a diversion, Australia could be a gainer, because it holds the world's largest reserves of high-grade uranium. But to most conservationists, global warming – while currently in the spotlight – is actually viewed as a lesser peril than an increased global emphasis on nuclear power. In their eyes, global warming is serious but not so serious as to merit a solution that might aggravate an even graver problem – a drastic increase in nuclear energy. The dilemma of fossil fuels is more complicated than is suggested by public debate within Australia.

Australia has come full circle in the past hundred years. At the start of the century it was self-sufficient in sources of energy to an astonishing degree. By 1950, the era of plenty had ended. The age of oil was here, but Australia produced no oil. Black coal was still vital but Australia, through poor industrial relations, failed to make the best use of it. By 1975, however, the nation was returning to self-sufficiency. Oil and natural gas had been found in large quantities; hydro-electric schemes had multiplied; brown coal was exploited massively and the mining of black coal was revitalised; and huge ships could carry coal cheaply from Australian ports on the Pacific coast to the booming markets in East Asia. By the year 2000, additional projects including the the natural gas of the north-west shelf in Western Australia had confirmed Australia's role as, overall, a massive exporter of energy.

The history of energy in this land has been marked by astonishing swings from abundance to scarcity, and from scarcity to abundance. It would be rash to think that another such swing is unlikely in the next fifty years. ■

Geoffrey Blainey has written many works on the history of Australian mining. The best known is *The Rush that Never Ended*, the first edition of which was published by Melbourne University Press in 1963.

Essay:
Beyond greed

Author:
Peter C. Doherty

Is this whole global warming scenario real or, as some newspaper columnists like to suggest, a massive conspiracy by self-serving scientists and self-appointed environmentalists who are trying to maximise their own resources, influence and power? Interestingly, we are starting to see both prominent political figures of the "right", and even some of the international energy companies, moving to the "left" of the more reactionary media on this issue. Maybe some of those organisations have the recent legal histories of the tobacco and asbestos industries in mind. Maybe they are also realising that they must diversify and adapt if they are to survive in the long term. After all, there can't be an infinite future in marketing a dwindling, natural resource. Other energy companies, though, are in denial and do their best to frustrate debate.

At least for the politicians, my guess is that they are reacting to a real shift in public perceptions of global warming. The Federal Environment Minister, Senator Ian Campbell, is an appealing personality who certainly "talks the talk", but we shall see in the longer term whether the emphasis of the administration he represents on voluntary controls will prove an effective way to "walk the walk".

Any newspaper editor will tell you: bad news sells. My sense is that many, if not most, of us are buying into the idea that global warming is real. Television presents us with an endless catalogue of disasters: the frogs are dying, the bushfires getting worse, reports of the hottest day on record.

The problem is not so much to convince people that we have a problem as to work out how to do something about it. Living in Memphis, Tennessee, I had to have my car exhaust checked annually at a municipal testing station. If your car doesn't pass, then it costs money to make it comply. Also in Memphis, the double-hung windows of our 1903 house were made more energy-efficient by the addition of external, triple-track storm windows fitted by the previous owner as part of a government-supported initiative. The triple glass/flyscreen windows represent a simple, relatively cheap and effective "bolt on" technology available in all big United States hardware stores, but I can't find them in Australia.

We could do a lot to conserve energy in the way we build and utilise our living spaces but, because it costs money, it will take a carrot-or-stick approach to make most of us react. Every individual has a part to play. We need to focus as much on "me" as "them and they" when it comes to climate change.

A little history: beginning less than 300 years ago with the Industrial Revolution, we have been releasing the stored energy of forests that compacted over millions of years with ever-increasing speed. The Earth itself is billions of years old, the adaptive immune system that I study first emerged in the bony fishes about 350 million years ago, humans (*Homo sapiens*) have been around for 100,000 to (at most) 200,000 years, we have had mining and agriculture for less than 10,000 years and cars for about a century.

The "prosperity" in terms of access to consumer durables, international travel and so forth that most middle-class Australians and Americans now enjoy certainly wasn't the reality for other than a very small minority only fifty years ago. There is nothing immutable about our current lifestyle, and no divine right that it can or should continue. Even so-called "conservative" politicians, though, can run enormous risks if they try to introduce just the smallest element of a reality check. Look at what the 1980 oil crunch did to US President Jimmy Carter who, as a born-again Christian from the American South, is hardly a radical.

President George W. Bush tried to make the point in his 2006 State of the Union address that it's past time for the US to kick its dependence on Middle East oil. Many of us had hoped to hear that from him immediately after September 11, 2001, but better late than never. His statement went down like a lead balloon. Conspicuous, mindless consumption of this non-renewable resource is broadly seen as an entitlement. The possibility that the couple of thousand American boys and girls from towns in rural "middle" America, the south and the Hispanic communities of the big cities who have died (many more have been maimed) in Iraq might in some way be connected to patterns of domestic oil consumption doesn't seem to have crossed into the wider US national consciousness.

The lesson is that, no matter how pervasive the global-warming argument, no matter how good the evidence, the only thing that will persuade many human beings to moderate their behaviour is to make environmentally damaging practices either expensive or illegal.

I'm not an authority on climate change and, though I'm a working experimental biologist, this is too complex an area for me to claim any authoritative position. My professional obsession is with understanding, and hopefully enhancing, immunity to the influenza A viruses. This has assumed much greater significance as we sit and watch the extremely dangerous H5N1 bird 'flu spreading throughout the world.

Of course, if the H5N1 viruses do jump the species barrier and kill off thirty to fifty per cent of the human population it would, at least for a time, diminish the population pressure that most consider a primary driver of global warming. The number of people on the planet has increased at least fourfold over the past hundred years, sixfold from the beginning of the Industrial Revolution.

Clearly, this rate of population growth, and perhaps the current global population size itself, is unsustainable. That reality still seems to escape some religious leaders, who continue to urge their followers to out-breed the competition. Fortunately, many of the faithful either have more sense, or are too financially constrained, to follow their directions. Some economists also seem to be in total denial about the possibility that human numbers cannot continue to grow exponentially.

As we apply the new genomic sciences to the study of human evolution, we are finding hints in our DNA history of genetic "bottlenecks" where the numbers of people were remarkably reduced. The "culling factor" may well have been infectious disease. Mortality rates of thirty to fifty per cent were recorded routinely in the plague that raged repeatedly through European communities in the middle ages. Currently, about three million people (the population of Melbourne) are dying annually of AIDS, but the human immunodeficiency virus (HIV) transmits at a relatively low rate and the effect on global population size is, so far, not very big.

People like me are dedicated to seeing that modern communities don't experience anything like the catastrophe of the medieval plague years. The effectiveness of the global response to the 2002–03 SARS epidemic is proof of this. Political systems, scientific expertise, business and regulatory authorities came together to protect the human community.

One might take the very harsh view that preserving human populations is counter-productive for the health of the planet, but it is only by assuring people in (particularly) the developing world that their children will survive that we can expect them to reduce family sizes. Stability, progress and good health go hand in hand.

The fact that warning bells about the dangers of global warming are being sounded loudly by all the national academies of science should cause us to think that we may be facing a substantial problem. The academy memberships are comprised largely of prominent, established scientists elected on the basis of achievement. As a consequence, they tend to be conservative, and try very hard to be responsible, to work effectively with their respective national governments and to be seen as the reservoir of informed scientific opinion. The most prestigious is probably the US National Academy of Sciences (NAS), which was established by President Abraham Lincoln exactly for those purposes.

The much older (1660) British national academy, the Royal Society (RS) of London, provides the model for the Australian Academy of Science (AAS), the organisation based around the igloo dome on the edge of the Australian National University campus in Canberra. The NAS, the RS and the AAS have all contributed, with other prestigious academies like the sciences section of the Académie Française, to issue joint reports on global climate change. In addition, each of these organisations has groups working continually on specific aspects of the problem. You can access much of this via Google, and the printed versions of their various reports are available for purchase or in good libraries. Most are readable, and don't require that you be a scientist to understand what is being said.

Beyond that, what is worth reading on this issue? There are some very committed Australian scientist communicators who focus on environmental issues, particularly Ian Lowe and Tim Flannery. Jared Diamond's *Collapse* (Viking, 2005) is, along with his *Guns, Germs and Steel* (Norton, 1997), a must-read for anyone who cares about the big themes of how human societies are shaped by, and shape, their environments.

Over the past three years, I've seen some good, in-depth, well-researched investigative articles on the environment, and other "off the top of the head" opinion pieces that both plumb the depths of intellectual dishonesty and show a profound, and arguably deliberate, ignorance of how science works and the nature of the world around us.

Television does a great job when it comes to conveying the acute reality of natural disasters and environmental catastrophes. But, in fact, some of the more spectacular horrors, like the Boxing Day tsunami, have nothing to do with either global warming or, as some of the more despicable clerics claimed, God's wrath, though they may be described as "acts of God" by the insurance industry. The fact is, tectonic plates moved, threw up a mountain under the ocean and caused a tidal wave.

It still isn't clear to me that global warming was a major player in the devastation of New Orleans by Hurricane Katrina. The incidence of the severe storms and hurricanes that can result when a low-pressure system comes in contact with a warmed ocean has been increasing dramatically in Florida and the Gulf of Mexico over the past decade or so. The next few years will no doubt tell us whether this is part of a continuing trend or just a consequence of some sort of climate blip that happens from time to time. A factor in evaluating this will be the development of ever more sophisticated computer hardware and software for environmental monitoring and prediction. The better and more comprehensive the data sets, the faster we can crunch the numbers, the more sophisticated our understanding and our capacity to deal both practically and intellectually with these very complex issues will become.

My bet is that much of the pressure to enhance these types of analytical tools will come from the insurance companies that stand to lose so much in many global-warming scenarios. By steadily increasing premiums, the insurance industry may also prove to be well ahead of both the political and the scientific communities when it comes to changing public behaviour that impacts on, and reacts to, global warming. That may already be happening in Florida.

Science is all about measurement, and numbers matter a great deal. What numbers should we look for?

The fact that atmospheric carbon dioxide levels have been rising rapidly since we started burning large quantities of fossil fuels at the beginning of the Industrial Revolution seems incontrovertible. Much of the "back-information" has come from the measurement of gas levels in, for example, air bubbles trapped in ice formations. Everyone is aware of this, and we should expect to continue seeing such numbers published.

The evidence that mean global water and air temperatures are increasing also seems valid, but this can be a very confusing area, even for the experts. My naïve perception is that cloud effects can cause confounding, and unpredicted, consequences. As the warming of the deep ocean proceeds, the "tractor" of the Gulf Stream that makes life more temperate for much of Britain and north-western Europe may stop, leading to a transient "ice age" in those areas.

A key factor to monitor for the oceans is evidence of species loss, particularly corals that are likely to bleach and die if mean water temperatures rise more than two degrees. We can expect that Australia's marine biologists will be watching this very closely.

The other parameter that affects the health of the oceans is acidity. Ocean acidity gives an objective measurement that is directly related to atmospheric carbon dioxide levels and thus human activity, which is why people who argue that global warming is a scam never mention acidity. Atmospheric carbon dioxide combines with water to give carbonic acid, a "weak" acid that in turn initiates further acidification pathways. We are familiar with this from acid-rain scenarios. The consequences can be disastrous for many ocean life forms, like corals, that need to make calcified shells. Again, Australia's marine biologists and climate scientists, working from the tropics to the Antarctic, are incredibly important monitors of this situation.

The melting of glaciers, the northern and southern polar icecaps and the Greenland icesheet also provides a spectrum of parameters than can readily be measured by, for example, satellite mapping. Again, as a society, we must ensure that this information continues to be freely available and, as individuals, we need to keep these numbers in our consciousness.

If you are a young person reading this, I would like to persuade you that one of the best things you can do is to spend at least a little of your time learning biology and some of the chemistry and physics that affect the environment. It's your future, and the future of the children who come after you, that we are talking about. Strength comes through knowledge and insight. It's great to have an emotional commitment to the environmental movement but it will mean a lot more to you and you will be much more effective if you can back up that commitment with at least some understanding of the underlying science. ■

Peter C. Doherty is the author of *The beginner's guide to winning the Nobel Prize,* (MUP, 2005), and Laureate Professor in the Department of Microbiology and Immunology at the University of Melbourne. His essay 'Beating the tyrant distance' was published in *Griffith REVIEW 6: Our Global Face*, Summer 2004 –05.

Essay:

Changing public attitudes to long-term issues

Author:

Ian Lowe

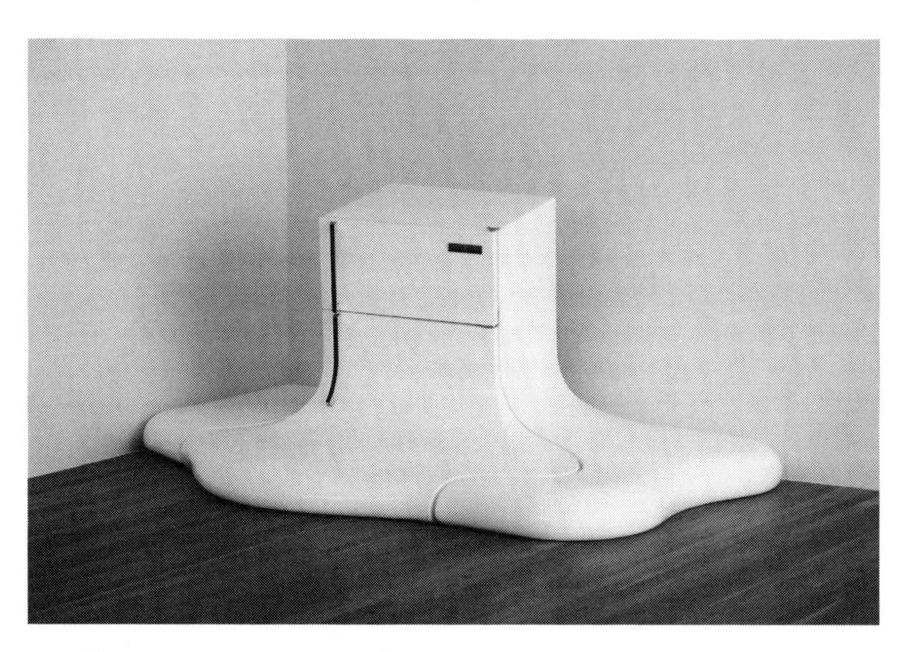

Image: Christian de Vietri / Einstein's refrigerator 2nd law 2004 / © Adrian Lambert 2004. Acorn Photo Agency.

Changing public attitudes to long-term issues

If industrial society is to survive, the next century will have to be a time of transformation, not just in technological capacity but also in our approach to the natural world and to each other. The second report in the Global Environmental Outlook series from the United Nations Environment Program says: "The present approach is not sustainable. Doing nothing is no longer an option."

A sustainable society would not be eroding its resource base, causing serious environmental damage or producing unacceptable social problems. But we are dissipating resources that future generations will need, damaging environmental systems and reducing social stability by widening the gap between rich and poor.

It is possible to move to a sustainable future, but it will require fundamental changes to our values and social institutions. While the UN report says that doing nothing about the huge problems we face is not an option, it remains the most common response of today's decision-makers; our National Strategy for Ecologically Sustainable Development, adopted in 1992, gathers dust in government pigeonholes. So how likely is it that we can achieve the fundamental changes needed for our civilisation to survive?

While sceptics point to the durability of cultural institutions and the reluctance of people to make voluntary sacrifices for the common good, we know that cultural traditions do change and people do make voluntary sacrifices. The examples that follow illustrate the willingness of people to make short-term concessions for the long-term good or to accept restrictions of their personal freedom for the good of the whole community. These examples suggest practical ways of achieving the sort of transition we will need for a sustainable future.

Perhaps the most dramatic example of recent change in our culture is the shift in the past forty years of attitudes to smoking tobacco, especially in shared spaces. In 1972, a persistent writer of letters to the existing domestic airlines celebrated the concession made by one of them to introduce a limited trial of setting aside a few rows of seats on some flights to be designated "non-smoking". Up until that point, it had been presumed smokers had a right to light up anywhere on aircraft, buses and trams; the same tolerance applied to restaurants, meeting rooms, public buildings and, in some states, even to cinemas.

The change on airlines was truly dramatic. With a choice, more and more customers asked for "non-smoking" seats until finally, in 1992, the Australian

Government announced a total smoking ban on domestic flights. There was barely a flicker of protest against the total prohibition of an activity that had been presumed acceptable only twenty years earlier.

In the workplace, the crucial event was the successful legal action by Liesel Scholem against the New South Wales Department of Health. Scholem presented medical evidence to show that her emphysema was a consequence of the department's allowing smokers to pollute the air she breathed in a shared office. Within a year of the verdict, many workplaces became non-smoking areas, forcing smokers to stand on the footpath outside their buildings. With the evidence accumulating that "passive smoking" is a health hazard, the prohibition has spread to most public-transport vehicles, public buildings, restaurants and even pubs. Most smokers have conceded that their right to smoke is overruled by the rights of others not to breathe exhaled smoke. As the right to smoke has been curtailed and more people have become aware of the health risks of smoking, the proportion of the adult population who smoke has fallen dramatically, from about half in 1950 to about a fifth today – a radical change, by any standard.

There are other, equally striking examples of radical changes in attitudes. When I first obtained my driving licence, police were allowed to stop a vehicle if they had reasonable grounds for believing the driver could be drunk. So a driver who attracted attention – for example by weaving all over the road or going through a red light – could be stopped, but an inconspicuous drunk could drive with impunity. Random breath tests were widely considered an infringement to personal freedom when first suggested in the 1960s. We are now accustomed to a regime in which drivers can be stopped "anywhere, any time" and be asked to provide a sample of their breath.

There is a close parallel with seatbelts in cars and helmets for motorbike riders. In each case, the compulsion was seen by some as an unacceptable intrusion, despite the overwhelming statistical evidence that the measures saved lives. Australia introduced compulsory seatbelts before most other nations. I recall bemused critics in the northern hemisphere seeing the measure as evidence we were sliding down the slippery slope to some sort of police state.

When I was young, plastic shopping bags did not exist and shoppers took their own permanent bags to shops. The plastic bag arrived and spread, so for a few decades there were very few shoppers taking their own bags to the supermarket. Now that increasing numbers of people are concerned about the wasteful practice of using throwaway bags, we are seeing a return to the practice of shoppers taking their permanent shopping bags. In some suburbs, those still using plastic bags are a small minority, attracting the attention and sometimes the opprobrium of other shoppers.

As a final local example, in the 1950s, soft-drink bottles incurred a deposit, so users returned them to get a refund. There was no deposit on beer bottles but community groups like the scouts collected them and were paid to return the bottles to distributors. We then passed through a throwaway era, with bottles used once and put into rubbish bins – a radical change. Recycling began in a half-hearted way, requiring trips to designated collection points, so few people took the trouble to return empty bottles. Then kerb-side recycling was introduced in many areas, with a consequent dramatic increase in the amount of glass, metal and plastic recycled. As with shopping bags, we passed through a stage of wasteful resource use before returning to the more responsible use pattern of fifty years ago: two opposite radical changes in less than the average human lifetime.

Probably the most striking example of radical change in the past twenty years is the end of communism, most dramatically symbolised by the fall of the Berlin Wall. For decades, the wall divided the city of Berlin and was patrolled by armed guards to prevent people fleeing gloomy East Berlin for the bright lights of the West. As a physical structure and as a symbol of the power of the communist government of the German Democratic Republic, it seemed absolutely impregnable. Only a handful of people succeeded in escaping to the West; many perished in the attempt. Then, almost overnight, the wall was abandoned. Joyful Berliners tore pieces out with their bare hands. Today, Berlin is a united city, once again the cultural and political centre of Germany. A small museum has preserved the history of the wall and entrepreneurs sell small fragments of the structure to tourists. But, for most visitors and residents, it is as if the barrier had never been there. The waters of history have closed over it.

The end of apartheid in South Africa is a similar story. For nearly a century after the Boer War, the Afrikaners ruled the country with their strange mix of Christianity and racism. The people of South Africa were divided into the three groups of whites, "coloureds" and Africans, a process that required some arbitrary decisions despite penal sanctions for sexual activity between members of different classifications. Public spaces such as beaches and even park benches were marked according to which group could use them. Those seeking change inside the country, like Nelson Mandela, were imprisoned or executed. Internal protests were ruthlessly suppressed in such actions as the infamous Sharpeville massacre. Anti-apartheid feelings in other countries became stronger, but conservative politicians sided with the regime, with the Queensland premier, in 1970, going to the ridiculous length of declaring a state of emergency to allow police to deal with those protesting against a Rugby tour. Sports people who refused to play against official South African teams were ostracised, in some cases ending their representative careers. Calls for economic sanctions or sporting boycotts were derided by business leaders, sporting officials and most politicians.

Eventually, the rising tide of political reaction forced politicians in many Western countries to support pressure on the South African regime, leading in time to the dismantling of apartheid and the enfranchisement of the entire adult population. Nelson Mandela, released and almost saintly in his willingness to forgive his tormentors, became president of the "rainbow nation" and the long-banned "Nkosi Sikelele Africa" was incorporated into the national anthem, with Mandela's wearing of the country's Rugby jersey at the World Cup final a powerful symbol of reconciliation. Many of those unable to face the dismantling of their system of privilege fled the country, but most remained and adapted to the new reality of twenty-first-century South Africa. While a hundred years of institutionalised disadvantage meant there was a need for some arbitrary measures to accelerate change, such as quotas in sporting teams, South African cricket and Rugby teams are increasing representative of the diversity of the country. As with the Berlin Wall, it is difficult now for a younger generation to imagine the old system even being defensible, let alone appearing unassailable.

There are more modest examples of radical changes in community attitudes enabling political reform. London Mayor Ken Livingstone decided to try to cut through the policy impasse that was making the city unworkable. With the public-transport system becoming steadily older and less attractive to commuters, more people were taking cars into the city, producing congestion and pollution. The reduced patronage of the public transport system meant it was unable to raise the revenue to invest in improvements. Livingstone implemented a "congestion charge" – essentially a levy on those bringing motor vehicles into the inner city area – to fund the revitalisation of public transport, arguing that sticks (financial penalties) and carrots (better public transport) were needed to change behaviour. The levy was not trivial, equivalent to about $12, and there were dire predictions of economic and social chaos before the introduction of the scheme.

Instead, it was so successful that, after the first year, Livingstone had widespread support for increasing the levy to the equivalent of $20 to speed up the raising of funds to improve the bus and underground rail systems. Far from creating chaos, the change has made the inner city much more pleasant – the streets are less congested and the air is cleaner, leading to an increased willingness of people to be in the city. Instead of the feared collapse of commerce, shops are doing more business as people return to the centre of the city. This was confidently expected by the planners, because it is exactly what has happened when streets have been closed to create shopping malls. Now that the change has been made, it would be politically very difficult to return to the old ways.

As I was finalising this article, Sweden announced its commitment to clean energy, with a goal of phasing out coal and oil by 2020. Iceland had set the same objective in 1999. Denmark, which already gets about a fifth of its electricity from wind power, aims to increase the level to half by 2020. Public opinion in the Scandinavian countries reflects increasing awareness of, and concern about, the two problems of "peak oil" and global climate change. This has made it not only possible, but politically wise, for governments to adopt energy policies that appear radical from the perspective of Australia, where politicians are still in denial about both.

It is clear from these examples that it is entirely possible for fundamental changes in public attitudes to happen. It is equally clear that it is no small task to achieve significant change at any level – the personal, the household, the small community or the nation. Most of the time, most people are reasonably content with the way things are and reluctant to embrace change, especially at a fundamental level. I believe that there are four steps to major change: discontent, a new vision, viable pathways and commitment.

Unless there is discontent, there is no motivation to change. If I am happy with my fitness, I am unlikely to change. If we are happy with our lifestyle, we are unlikely to change – which is why advertising tries to make us discontented and therefore more susceptible to the latest consumer fad. Richard Eckersley argues that the seven deadly sins – envy, greed, lust, pride, anger, sloth and gluttony – have been re-packaged as the seven marketing imperatives of the modern world. They are constantly used, as Clive Hamilton puts it, to urge us to spend money we don't have to buy things we don't need to impress people we don't like. Producing dissatisfaction is the aim of all advertising.

I am dissatisfied with our lifestyle because we are depriving future Australians of the sorts of opportunities we have taken for granted. Unless that awareness is widespread, there will not be the dissatisfaction that will motivate change.

Being discontented is not enough. Even if I am unhappy with my fitness or my lifestyle, I won't do anything unless I can imagine a better alternative. Without a vision, change might actually make things worse rather than better. So the second step to real change is to picture the future we want. Advertisers use this principle, portraying the life of indolence or the sexual pleasures that await us if we buy their products. In the case of my fitness, I can imagine the sort of state I would like to achieve, because I was there a few decades ago.

Achieving a sustainable society is more difficult because we haven't been there, so we have to invent it. Unless we have a credible vision of a future sustainable society, it is very unlikely there will be support for change. In writing *A Big Fix* (Black Inc, 2005), I described the essential components of a future sustainable society. There is room for argument about some of the details, but the general principle is clear. We must limit human consumption so it doesn't exceed the sustainable level of production from natural systems.

Again, even a coherent vision is not enough if we can't see a way of getting there from where we are. I do a fair bit of bushwalking, so I know that the easiest route from your starting point to your destination is hardly ever a straight line; you need to take account of the lie of the land and the vegetation. A friend might advise me against trying to recover my fitness of twenty years ago from where I am, but I might see a viable pathway, with suitable attention to diet and exercise. So the third component of change is developing and describing feasible pathways that would take us to our goal from where we are now. Again, it is more difficult with a future sustainable society. Even when a future sustainable society can be imagined and described, it is difficult – and contentious – to spell out the steps that will take us there.

Since a sustainable future must involve much less use of fossil fuels like coal and oil, the initial steps run counter to some very powerful vested interests. In Australia's case, the export coal industry and the heavily subsidised aluminium industry have been remarkably effective in blocking the sorts of changes that are clearly in the public interest. Before the Kyoto conference on climate change, which set preliminary targets for reducing the levels of greenhouse pollution from the industrialised world, I attended a briefing for interest groups. I sat with mounting incredulity as the leader of the Australian delegation set out the stance being proposed for the Kyoto meeting. When I asked whether it was possible to identify any nuance in the official position that differentiated it from the public stance of the energy-intensive industries, the delegation leader responded angrily, but was eventually forced to concede that there was no significant difference. The position adopted at Kyoto by the Australian Government, led by the Minister for the Environment, Senator Robert Hill, was to defend the interests of two industries that are largely foreign-owned – the export coal and aluminium businesses. Another example of a vested interest that has been very successful in blocking change is the road-freight operation, heavily subsidised because the registration charges and fuel taxes paid by large trucks are thousands of dollars less than the damage they do to roads. These examples illustrate that the pathways to a sustainable future must be politically feasible as well as technically viable.

The fourth component is commitment. If I can spell out a coherent program of diet and exercise that will enable me to regain my physical fitness, it will still take commitment to stick to the regime. We all know people who have good intentions but lack the will to stick to their plans. The politics of sustainability hinges on developing such a public commitment to a sustainable future that the change becomes politically irresistible.

As these examples show, traditional politics will only respond to urgent long-term issues when there is irresistible pressure. The historian Paul Kennedy argues that those who succeed in democratic political systems are usually those who have refined to an art form the avoidance of any threat to powerful interest groups. So they are unlikely, he proposes, to adopt significant changes in the interest of future generations as long as they can argue that experts are divided or more research is needed.

In the case of complex issues like global climate change, there will always be some division among experts, making it feasible to claim that more research is needed before we embrace significant change. It has become steadily more difficult to find scientists with any credible claim to expertise who will deny that human activity is changing the global climate, but media organisations and governments have responded by intensifying the search rather than accepting the overwhelming consensus.

While several European countries have now adopted serious targets for reducing greenhouse pollution, in Australia we are still getting evasion and spin from our national government, like the extraordinary claim that we are "on track to meet our Kyoto obligation". The basis for the misleading claim is Australia's uniquely generous target, an eight per cent increase in emissions when most developed countries are required to reduce, combined with what is known around the world as "the Australia clause" in the Kyoto Protocol. Our delegation demanded that changes in land use should be counted in the greenhouse-pollution calculation. So our baseline was inflated by the huge scale of land clearing in 1990, and the Beattie Government's actions to stop that clearing allows the national government to claim we are doing our share for the global effort. In fact, our energy-related emissions, mainly from electricity and transport, are more than a third above the 1990 Kyoto baseline and spiralling out of control, with no serious policies even to slow down the rate of increase.

The crucial first step will be committing to the principle of a sustainable future. It should not be necessary to say this, because the Council of Australian Governments adopted the National Strategy for Ecologically Sustainable Development as long ago as 1992, but there is little sign of these ideals in 2005. We should be flexible about how we get there, but the goal must be unambiguous.

In *A Big Fix*, I tried to set out the the principles of a future sustainable society and a process for achieving that goal. This year, the Australian Conservation Foundation has set up a group called Australia's Future Makers. The aim is to develop a charter for a sustainable future and gather public support from prominent Australians, leading up to a Sustainability Summit in October. It is intended to create a groundswell of public opinion that politicians will be unable to ignore.

The crucial step will be to develop awareness that the change will benefit all of us, rather than being a massive sacrifice in the short term to help future generations. It will then be feasible to advise voters at future elections on the policies that are needed, allowing them to make their own judgements about which candidates or parties will be more likely to fulfil our hopes.

It is sometimes claimed that environmentalists are trying to "spoil the party" and stop people enjoying our modern lifestyle. Rather than spoiling the party, I am suggesting that a new party is starting up. It is a better party because it won't run out of food and drink, it will be more satisfying because it will be based on personal fulfilment rather than gluttonous consumption, it won't damage our shared "house" or leave us with a hangover, and it won't have envious neighbours looking on or throwing rocks on the roof. It will be a party we can all enjoy, and a party we can expect our children to enjoy as well. All around the world, people are striving to develop social and institutional responses that will enable the transition to a sustainable future. We have to recognise above all else that we share the Earth with all other species and hold it in trust for all future generations. That moral duty provides the impetus for the transition to a sustainable future, just as moral principles underpinned the end of apartheid and the dismantling of the Berlin Wall. As in those cases, the changes we need are not inevitable, but those examples serve as a reminder that radical changes happen when enough of us want them.

So what are the realistic prospects for the changes we need to make? Some appear relatively straightforward. For example, I think there would be widespread public support today for a responsible approach of phasing out coal-fired power as existing plant reaches the end of its life, replacing it with a mix of renewable fuels. There would be little public objection to improving the efficiency of appliances and vehicles, since this provides economic gains as well as reduced environmental impact. Replacing dangerous pesticides with safer alternatives should not be contentious, while there would probably also be political support for preferring local foods to those carted halfway round the world.

The really problematic changes relate to growth in population and consumption. While it is clear from elementary biology that no species can increase its population without limit in a closed system, most of us are still in denial about the application of this iron law to humans. In similar terms, our right to consume more per person is almost the defining myth of our time. The real challenge is to spread awareness of the need to achieve a future steady state and develop political support for that change. ■

Ian Lowe is an emeritus professor at Griffith University, the author of *A Big Fix* (Black Inc, 2005) and President of the Australian Conservation Foundation. His essay 'Radically rethinking a sustainable future' was published in *Griffith REVIEW 2: Dreams of Land*, Summer 2003 – 2004.

Fiction:
The future from the bottom of a boat
Author:
Andrew Belk

Image: Ken Searle / Cooks River, High Noon 2002 / oil on hardboard 122 x 92cm / Courtesy the artist and Watters Gallery
Photograph: Michele Brouet

The future from the bottom of a boat

The houses in the estate are all the same. Low-slung brick with dry brown roofs and yards. The fences are thin fibro planks woven slackly between raw concrete uprights. Deb doesn't trust the fence. It's like someone tried too hard to make something out of nothing.

All is rust. It flows down the walls from gutters and window frames and vents. It streaks from fastening bolts down the fence planks and piles up on the earth. It makes holes in cars. It lives on the spokes of pushbike wheels and arrives overnight on the steel lock tabs of school cases. It is your point of difference. It is how you know where you live – what is yours. Here, decay is personal.

Deb stands in the middle of the cul-de-sac and licks salt air from her lips. Tomorrow she will start Year Three. Her sister drops a canvas bag at her feet and bends to knot the drawstring. Tomorrow, Tracey must choose between starting Year Eleven and taking a job at Collie 3 power station. Either way, she figures, she loses. She sees Deb's painted toes poking from her small plastic sandals. Yesterday she hit her little sister in the head for using her nail polish. She reminds herself not to be a bitch. Tells herself Deb is still a baby. She says, "Let's go."

They have the bus mostly to themselves as it works the interlocked points and bays of the southern backwater. A few workers get on and off at the power stations – the big coal burners slumped at the foreshore like dogs at a puddle, lapping water in cool and pissing it out steaming. The cooling inlets have red signs warning not to swim there. Some kid did once. He got sucked through the turbines while spearfishing for flounder and turned the outlet water crimson.

Tracey pushes her head against the glass and lets the vibration massage her temple. Her father has worked at the Collie 3 power station for thirty-two years. He opens and closes the valves that regulate the pressure on boiler number four. Her mother has worked there nineteen years. She cleans the administration buildings, the canteen and toilets. Tracey's job won't be like that. Her parents have called in their favours. "You can be someone," they tell her. "You can be in admin." As the bus passes the foaming outlet, Tracey scans the water for traces of her own blood.

Deb is watching the outlet, too. Nobody fishes there anymore, scared they might accidentally eat a fish that has eaten some of the turbine kid. Deb has a term for kids like him – Example. Example is a kid who has something terrible happen to them, and because of that terrible thing happening, causes a change. The turbine kid is Example because after he got chopped to burley, the power stations put mesh around their inlets. There is the kid who's the Example of why you have to wear a bike helmet. The kid who's the Example of why brake-fluid bottles have childproof caps. The kid who's the Example of why swing seats are made of plastic and not hardwood. The kid who's the Example of why pool filters have covers. There are lots of Examples, and as far as Deb can see, they are usually kids from cul-de-sacs.

At the approach to the northern waters, the driver parks the bus on the verge and changes the route indicator from 363 to 17. The sisters watch as he pulls up his socks, tucks in his shirt and checks his nostrils in the long side mirror. The 17 now travels the green, clear-water suburbs to the city. *But it's the same bus*, thinks Tracey. She thinks of how, on the worn brown vinyl seats, sweet-smelling office workers and dressed-up shoppers swap dead skin and sweat with the dirty bums of power-station workers and housing-estate kids. She thinks how every morning a little coaldust or algal bloom makes its way to the city and how every evening a little fine perfume or imported silk makes its way to the estate. Tracey knows it means something.

A small sheltered bay opens beside them. The water is soft and clear and buoyed with yachts. Deb presses her face on to the glass next to her sister and asks if this is the place. Tracey says "Yes." She lifts Deb high enough to pull the bell cord and the little girl smiles her first smile of the day. They stand on the roadside and watch the number 17 diminish in a veneer of diesel. *But it's the same bus*, thinks Deb.

Inside the rotunda of the Rotary Park the girls unpack their canvas bag. They eat tomato-sauce sandwiches. They drink lemon make-up cordial. They put together fishing rods, silently pulling soft green line through golden swivel eyes. Tracey tells Deb to remember that if anyone asks anything, to shut up and let her do the talking. Deb tells Tracey she doesn't have to keep on telling her.

The pair walk the park foreshore, flicking and retrieving their lines, lazy silhouettes before the sunset. At the park's edge they come to the first of the waterfront houses. Tracey remembers when there were fishing shacks and half-arsed weekenders shufffled on this shore. Now a vista of anal retention sits tight above the retaining wall. The narrow strip of leftover sand and seaweed between this wall and high tide forms the public right of way. The girls walk this strip, flicking and retrieving, pretending not to look. Deb has never been this close to the ocean inlet, never seen the water so clear. A baby flathead falls for her wobbler. She unhooks it lightly and gives it back to the lake.

Deb watches the flathead scoot into the blackening water and hopes it doesn't die. She hears her sister swear beside her and flinches, waiting for a smack in the head. Tracey has told her three times they aren't there to catch fish. When the blow doesn't come she looks up to see their way blocked by barbed wire cutting across the path and running out into the water. On the other side is a mansion of unlikely angles and windows that cannot be opened. It reminds Deb of their fence.

Tracey swears again. "This is why we are doing this," she says. She tells Deb the fences are meant to stop at the retaining wall. She tells her people are meant to be able to walk all the way around the lakefront if they want and how some woman did once and it took her eighteen days. She tells her how the council told the rich people it was illegal to put fences down into the waterline but the rich people did it anyway. She tells her how the council threatened to sue the rich people, so the rich people became the council and that was the end of that.

The last of the day falls into the water. Deb thinks about sunsets and how people are mostly inside when they happen. She thinks about the time her visiting aunty went mental photographing one – saying over and over how she couldn't wait to show her friends in Melbourne. When the photos came back and were just bits of paper a billionth the size and feeling of the real sky, Deb had to bite her hand to hide her happiness. Tracey thinks about boiler number four and urinal cakes. She is lost on the far side of the lake, where red aircraft-warning lights on the smokestacks of Collie 3 glow below the evening star. It is darker there. She thinks about the coaldust blanketing the cul-de-sac, tucking it in, putting it to sleep. She walks into the lake.

The water is cold already. It rushes Tracey's jeans and chucks shivers up the quick of her teeth. By the time she rounds the fence it is above her waist. She lets go of a long defiant piss. It spreads over her crutch and down her thighs, briefly warming her. Behind, Deb is up to her chest, struggling to keep the balance between walking and floating. She doesn't like being in the lake at night. Unlike creatures of wardrobes and under beds, the fangs and claws and stings of creatures of the lake are real.

Tracey takes the hacksaw from the canvas bag and starts work on the chain securing the punt. Deb thinks it is the shittiest boat she has ever seen. She is seething – on the brink of storming off. All summer she has watched other people having a better time than her. Watched as they caught fish that belonged to her. Watched as they dived into water that belonged to her. Watched as they motored across long slapping waves that belonged to her. Without a boat it seemed everything good lay beyond. When it became clear the annual promise had been gambled, drunk and smoked again, the girls spent the rest of Christmas Day sitting in the shallows with their backs to the shore and the warm water up to their top lips. They breathed through flared nostrils. They looked out silently from frowns drawn hard against the glare. The shore breeze blew the peeling skin from their necks but it couldn't cool them. The notion that a boat should be stolen was floated. Later, it bobbed back. Justified.

The punt is three metres long at the most and pocked with crusty wads of fibreglass repair that lift and hang like ripe school-day scabs. Deep cracks down its sides are packed with rivers of crumbling sealant. The length of garden hose meant to protect its gunnel is brittle and fractured. There is no motor, just marks where it should be. Tracey feels Deb's mood. She knows if they walk this shore they will probably find something decent: maybe a four-metre tinny with a six-horse outboard like they talked of. Something big enough to get the two of them up and planing but small enough not to need a licence or registration.

Tracey tells her little sister that the pricks and their illegal fence need to be taught a lesson. She tells her, this way, they don't have to feel bad about stealing because it is just making things fair. She tells her because the rich people break the law, it is OK for them to break the law as well. She tells her that in the morning when these pricks with their illegal fence want to get out to their big flash yacht, they will have to swim. She says they will repair the cracks and dents and sand it back and paint it. She says they could paint it white on the outside and green on the inside.

Deb is quiet. Then she says, "Maybe we can paint it yellow."

The girls pull the boat into the lake. With one leg in and one leg out, Tracey pushes them off in a series of long bouncing strides. When the bottom drops away she lifts her leg high behind her and shakes off the water and weed. Crouching at the back of the shitty little punt like a figure skater, the smell of her sister's hair flowing through her, she is caught off guard by happiness.

Without oars, they lay over the bow and dog paddle the punt out through the anchored yachts. They whisper the names and ports of origin of each as they pass. Deb says, "Ella Angel, Vancouver." She says they should name their boat "Freedom" because they got it for free. Tracey thinks about all the water that passed under the Ella Angel as she crossed the world to get to their lake. She pictures a wave gently passing underneath the long white hull. She pictures the wave continuing across the ocean, carrying whales and dolphins and fish and turtles. She watches it duck the hard steel of a black coal ship then gently caress the rotting wood of a refugee boat. She watches it enter through the clean inlet waters and travel across the lake towards the shallows behind the estate. She sees the wave relax, eyeing a gentle shore break, lining up a pudgy boy in an inner tube for one last fling. She watches the cooling inlet of Collie 3 reach out and grab it. Drag it through the turbines and spit it through the outlet. Devoured. Ragged. Drowned.

Tracey lines up the bow with the green and red flashing buoys that mark the channel. She knows the tide is running in. If she can get the punt into its flow it will carry them across the lake. Even though the bus trip along the spindles of the shore is fifty kilometres, in a straight line it is less than ten. They can drift that in the three or four hours before the tide turns. Once they bump ashore they can walk the boat along the edge, or maybe come back later with some oars. Tracey tells Deb to sit up the back. She leans long over the bow and digs her arms deep into the cold black water. With Deb holding her ankles, she pulls them forward with strong butterfly strokes. When she eventually feels the overwhelming pull of the channel she lines up the lights of the Collie 3 smokestacks and lets the lake take over.

The girls find the most comfortable position is lying on the floor with their heads at opposite ends and their legs resting on the middle seat. They take it in turns to sit up and check they are drifting toward the stacks – at first every few minutes but now, trusting the drift is good, only occasionally. Low down inside the boat, her promise to her sister kept, the lake bed safely rising and falling below, Tracey dares to think about tomorrow.

She doesn't want to go back to school. She doesn't care about the assassination of Francis Ferdinand or the volume of a cone. She doesn't want to sit all day on plastic chairs, rote memorising what teachers copy from books. She doesn't want the constant hard invitations of boys driven dog wild by puberty. She doesn't want to work at Collie 3. She doesn't want to be in admin. She doesn't want to walk between those filthy stacks each morning. She doesn't want to get off the bus while it is still the 363.

Tracey knows she is supposed to want something but doesn't know what. Her only inkling comes from a foreman at Collie 3. Yesterday, he had given her a lift in a new car with a light top and dark windows. He told her he had bought it for himself as a thirtieth birthday present. He told her that was because he was making thirty thousand dollars a year. He told her not many blokes did that – made thirty thousand a year by the time they were thirty. "I could buy and sell your father," he told her. "The Deville is the biggest car you can buy," he said.

He bet her five bucks that if she lay on the back seat and stretched out she wouldn't be able to touch the sides.

Tracey thinks about how she is halfway to thirty. She thinks about how rust from the gutter bleeds the word "ill" onto the bricks above her bedroom window. She thinks about her father's swollen hands turning valves and her mother's blue-vein legs hosing shit. She says, "By the time I am thirty I am going to be making thirty thousand dollars a year."

Deb tries to picture her sister in fifteen years. It will be the year 2000. It is further than she can imagine. Deb doesn't like to think about the future. As far as she can tell, it is already history. "Probably the world will end before then anyway," she says.

Tracey's small happiness falls from her. It sinks to the bottom of the lake and is pulled apart by crabs. Like her, Deb is smart – a dangerous thing in a place where being too clever will get you smashed in the face as routinely as being too dumb. Tracey has watched with sadness this summer as Deb has come to know the truth. It has fallen to the big sister to help the little sister make the transition from bliss. It has been Tracey who has nursed Deb through nightmares of skeleton babies starving to death – their mother cleaning by night when offices are abandoned and reeking toilets are vacant. It is Tracey who has soothed the harsh screams of atomic bomb blasts – their father asleep from television and beer by nine o'clock. It is Tracey who has cradled the eight-year-old and kissed her back to sleep after frantic questionings about black smoke from the Collie stacks and global warming and the lake rising to take them while they sleep.

Tracey tells Deb the world isn't going to end. She tells her that by the year 2000, society will be very highly advanced. So advanced babies will not starve to death. So advanced atomic bombs will be dismantled. So advanced there will be no black smoke. She tells her that in the year 2000 society will be so highly advanced little girls like her won't have heard of starvation , won't have nightmares about pollution, won't know the terror of thinking they can be blown apart. She tells her that in the year 2000 the estates will have fences you can trust.

Deb listens as her sister gently talks of the future. A world of peace and happiness and robot dogs. Soon there is only the sound of Tracey's breathing and the soft lapping of waves on the hull.

Deb is too excited to sleep. Tomorrow she will start Year Three. She sits up and checks the drift. Her red hair has turned wild from the salt air and sits huge and bouffant on her small freckled head. The drift is good. They are headed for the stacks. ■

Andrew Belk is a Melbourne-based writer. His story, 'The Einstein canaries' was published *in Griffith REVIEW 9: Up North*, Spring 2005.

Legacy | tuvalu

Photographer Jocelyn Carlin

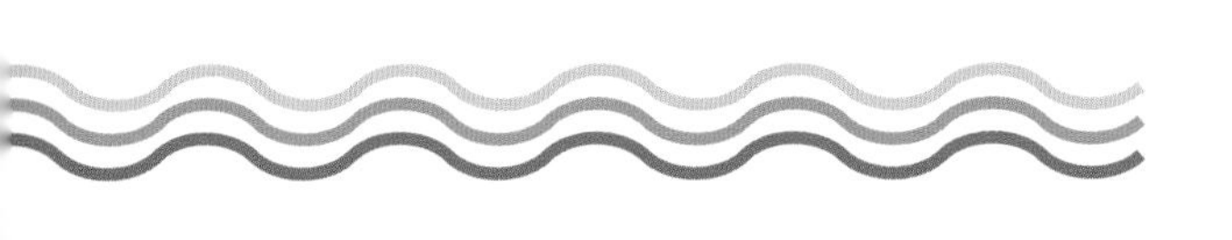

Reportage: **We are all Tuvaluans**

Reporter: **Mark Hayes**

Image: Tuvaluan high school students ponder their future / Photograph: Jocelyn Carlin

We are all Tuvaluans

T*atou ne Tuvalu Katoa* – "We are all Tuvaluans" is often used in Tuvalu as an expression of national unity, calling on islanders to pull together in the collective interests of their tiny, isolated and vulnerable country. It is also used by some environmentalists who understand that global warming and rising sea levels, while gravely threatening the existence of low-lying tropical island countries like Tuvalu, threaten us all.

Wednesday, February 22, 2006. It's going to be another hot and steamy day on Funafuti, the main atoll in Tuvalu. There is just the hint of a breeze, and the vague promise of a shower from the clouds to the north-east. This won't cool things down, just add to the humidity.

I'm standing in the middle of the widest part of Funafuti Atoll, near the airstrip that dominates this part of the island. Directly to the east, about 400 metres away, the coral rock and sand wall thrown up by a cyclone in late 1972 protects the atoll from surges. If it weren't for the breezy drone of the power station, I could hear the Pacific Ocean, rolling white coral rocks like bowling balls up and down the rocky, barren shore. I won't go out there unless I have to. It is scary and weird. You sense the brooding ocean is out to get the place.

To the west, up a paved lane, past the blue-roofed and white-walled three-storey Government Building, I can see the lagoon and, far away, the *motu* (islet) of Tepuka about twelve kilometres across *Te Namo* (the lagoon). *Te Namo* is placid this morning, its waves gently lapping the atoll's western shore.

Funafuti Atoll, viewed from space, looks like a boomerang. From the cockpit of the Air Fiji twin-engine, turbo-prop plane that makes the thrice-weekly two-and-a-half hour flight north to the tiny airport with the best destination code on the planet, FUN, Funafuti comes into view like a thin green snake laid almost south to north in the vast open ocean. It's twelve kilometres long, with *motu* at each end, and scattered along the western edge of the wide *Te Namo*. The atoll is about eight degrees south of the equator. There are no hills or mountains here, no rivers or permanent streams.

In 1897, the Australian scientist T.W. Edgeworth David led an expedition to Funafuti Atoll, then part of the British Gilbert and Ellice Islands colony – Polynesian Tuvalu was the southern, Ellice Islands; the northern Micronesian Gilberts is now Kiribati. Professor David wanted to test Charles Darwin's hypothesis that tropical atolls were perched and growing atop ancient volcanoes or mountains, so his team drilled holes deep into the centre of the atoll to partially confirm Darwin's theory. The nondescript spot – a small concrete circle with a hole in the middle tucked away near the hospital on a side road – is still called David's Drill or David's Hole, but I call it Mt Funafuti, much to the amusement of my Tuvaluan friends.

Tuvalu consists of nine low-lying islands with a total land area of twenty-six square kilometres. The name, taken at separation from the Gilbert Islands in 1975, means "eight standing together". The southernmost atoll in the group, Niulakita, has been re-inhabited more recently.

The national slogan, *Tuvalu mo te Atua*, means "Tuvalu for God", reflecting the islanders' strong Christian beliefs, with virtually all the population members of the *Ekalesia Kelisiano Tuvalu* (EKT), a protestant denomination based on the work of the London Missionary Society in the late nineteenth and early twentieth centuries.

The country has an estimated population of 11,500; just under half live on densely inhabited Funafuti, with a land area of less than three square kilometres. The problem of "urban drift" from the outer islands to the capital, increasing its population by a tenth in a decade, is significantly contributing to the country's challenges. Their home islands and extended family ties are central to Tuvaluans' identity, and so many people from the outer islands crowding Funafuti adds still more pressures on this atoll's environment and social fabric.

In every respect, Tuvaluan life is lived pretty close to the edge; either subsistence fishing, gathering natural or small plantation coconut, pandanus or breadfruit farming, or, more usually, a mixture of developing island Western-style living and traditional practices. The biggest employer is the Tuvaluan Government, the next biggest being a seafarer hire company that contracts 400 to 600 Tuvaluan men to crew cargo boats worldwide. Their remittances home help their families to live comfortable lives, but at the cost of long separations from wives and growing children. The average Tuvaluan family lives on about $1,200 a year, though many get more and some less.

This Wednesday morning, locals are starting to stir. The few who were sleeping on the airstrip have rolled up their *taapa* (woven coconut or pandanus fibre) mats and pillows and wandered back to their *fales* (open-sided, roofed, raised platforms) outside their houses for a little more sleep. They're not supposed to sleep on the airstrip, but it's one of the few places to catch a breeze on hot, still, tropical nights.

Solo, a man in his early twenties, walks towards me with a wicked-looking knife and a couple of bottles with woven string ties at their necks, smiles, bids me a cheery *talofa* ("hello") and starts climbing a coconut tree.

Along the road, locals are walking, or riding bicycles or small motorbikes. Children, some already in their crisp light-blue and white school uniforms, have been sent to the *fusi* (small co-op supermarket). Some folks are carrying large white plastic biscuit buckets filled with kitchen scraps and coconut meat and making their way to their pig pens along the seawall to the east.

High up in the coconut tree, Solo's busy, carefully paring back a frond stalk so the sap can flow better into the bottle he ties beneath to catch it. Two half-full bottles collected overnight are ready to be carried down. Locals use the sap, called toddy, to make a sweet dessert, a local liquor and for cooking.

Solo is whistling a rather tuneless song, and occasionally breaks into words.

"What are you singing?" I shout up at him.

"Oh, just a song," he says, in his quiet Tuvaluan way.

"What's it about?"

"It's a song to call the maidens," he grins.

I wonder what his young wife in their nearby house would make of this. Folks are passing as the day begins and we exchange waves or greetings.

The sun finally punches through the clouds to the east, bathing the atoll in slivers of bright white light, and I feel sweat starting to form on my back.

On the far side of the square, near the entrance to the Government Building, some men in overalls are sweeping up leaves and litter while a very bored policeman sits nearby watching. These are prisoners from Funafuti's gaol serving time for committing some of the few crimes serious enough to warrant custodial sentences. Most crime here is alcohol-related, and the worst offenders are unlicensed, often young, male drivers caught drink-driving.

Another challenge for Funafuti, since the roads were paved in 2002 using windfall money from leasing Tuvalu's dot tv internet country code, is the steady increase in vehicles. Not that there's anywhere much to drive to on an atoll this small, but locals seem to manage. More vehicles add to Funafuti's problems because, in the salt-saturated air, rust eats into the chassis, and abandoned cars and trucks litter the atoll, with some being used as part of the defences against sea surges from the lagoon.

Sema, a Fijian woman married to a man from Nukufetau, one of Tuvalu's outer islands, who works at the Filamona Lodge where I'm staying, comes to work, sees me writing, comes over and peers at the laptop screen. I want to get this just right, so I show my developing story to some locals. I'm a wordy *palagi* (white person, outsider) who's here to try to tell part of the current Tuvalu story.

"Oh, wow," Sema says. "That's good."

"*Fakafeti lasi*," I gratefully say, and return to the keyboard.

A bit later, Hilia Vavae comes by and also looks at my writing. She is the director of the Tuvalu Meteorological Office, housed in a white bungalow next to the power station on the eastern side of the atoll, and the local expert on Tuvaluan weather, with twenty years' experience. Her daily data is sent worldwide and added to three-day regional weather forecasts, and the fearsome climate calculations and simulations that fuel the scientific debates about global warming. She's happy too, so I'm satisfied I'm getting this right.

In six days, the highest predicted tide to hit Tuvalu in almost fifteen years will cause widespread local flooding. If you believed some reports about the country, locals should be huddled in fearful dread, counting the days until their beloved islands sink forever beneath the rising seas. This reporting drives my Tuvaluan friends nuts. "Parachute journalists," they mutter when this hype is published, raising their eyes upward in mixed annoyance and resignation.

Nobody's huddling in fear and dread. There are just Tuvaluans going about their everyday business.

Earlier this week, Radio Australia and ABC News Online reported the December 2005 findings from the South Pacific Sea Level and Climate Monitoring Project that showed that the seas around Tuvalu had risen about seven centimetres over the past thirteen years. The findings were hedged with scientific caveats, as these reports always are.

Seven centimetres' rise might not seem like much, but the highest point of land on this atoll is just 3.7 metres above mean high tide. The difference between the highest points of land here, which might be Mt Funafuti or a spot at one end of the airstrip, and the predicted high tide next week is the distance between my sandals and my knees.

With better internet access, and satellite television and radio being more common, Tuvaluans and visitors like me can more easily access the same news and online information taken for granted elsewhere.

High tides have been hitting these islands for far longer than the 2,000 years they have been inhabited. Coral atolls have been rising and disappearing for aeons. But scientific data and stories elders tell about the climate in living and oral memory, around which many island traditions are based, point to something strange – even ominous – occurring here and on other low-lying island countries.

If Tuvaluans are praying about the weather at all, they're praying there won't be a storm or a cyclone far away sending surges in this direction when the tide peaks. Tuvalu is outside "cyclone alley" in the south-western Pacific, and so the cyclones that form, menace and occasionally devastate Fiji, Samoa and Tonga to the south rarely hit. The worst cyclone to strike Tuvalu in living memory was Bebe in October 1972, which almost destroyed Funafuti. The country was directly affected, but not hit, by three cyclones in 1997. The worst casualty was a small *motu* called Tepuka sa Vili Vili out on the western edge of the lagoon, once capped with dense vegetation and ringed by dazzling coral sand, now a barren wave-assaulted brown rock.

Just past the southern end of the airstrip, in August 2002, a storm surge with three huge waves came out of the Pacific without warning and carved a swathe into the atoll 500 metres wide. The area was scoured back to coral rock, vegetation swept into tangled heaps. The video of the waves is shocking and locals who witnessed it were terrified. Luckily, nobody was injured and no houses or *fales* were damaged. When I saw the damage four months later, I was shocked. But the ocean always spooks me as it incessantly batters, nibbles and sucks at the atoll.

Yesterday morning, I walked along the airstrip and back up Tuvalu Road, greeting locals on the street and in their gardens. A teenager straddled a row of cabbage seedlings, dipping a tin can into a large bucket and carefully watering the plants while a gaggle of laughing children watched me watching him. His mother proudly showed me the cucumber vines, with bright yellow flowers, wrapping themselves over the trellis in front of their door. Next door, some green and yellow tomatoes were peeping from the neighbour's vines. Along the road, locals were carefully sweeping up leaves and fronds from the pandanus, breadfruit and coconut palms, and putting them in bags to use as mulch on their own gardens. Almost every house had a garden, something I had not seen during my last visit here late in 2004.

The sea is not the only threat to Tuvaluans, and this is part of the country's survival strategy. The diet, heavy with imported foods supplemented by seafood, local produce like pig meat, coconuts, breadfruit, taro and a large, slow-growing tuber called *pulaka*, contributes to the high incidence of obesity, hypertension and diabetes. Now vegetables are more abundant, and locally grown, the product of a Taiwanese-funded demonstration nursery and garden from which the government distributes seeds and information on gardening.

But they struggle to grow their version of potatoes, *pulaka*, as successfully as they once could, for reasons I'll explain.

Yesterday afternoon, after school, the village of Vaikau seemed to be awash with children, playing, riding bikes and filling the place with their squeals and laughter. On the veranda outside the church, a pastor was leading some kids in a Bible study, all hunched over their large and well-thumbed scriptures, and next door, in a *maneapa* (open-sided meeting house), smaller kids were amusing themselves while a Tuvaluan matron tried to keep their attention long enough to get a few "good words" into their young heads – without too much success, it seemed to me.

Keeping Tuvalu's strong Christian practices alive is also part of their survival strategy. They start very young.

Later, in the slowly gathering twilight, the whole airstrip, from end to end, over a kilometre, was full of people. A vigorous soccer match was underway opposite the prime minister's residence, then a touch football game, then some public servants from rival departments hard at volleyball. And everywhere, the laughing children.

This week, there's a workshop on the Tuvaluan National Adaptation Plan of Action (NAPA), a local version of a global program to help especially vulnerable countries respond to the potentially terminal challenges they face. Last week, the World Wildlife Fund Pacific led a workshop presenting the global and regional view of global warming and its effects in the Pacific. This week's workshop adds local perspectives. Representatives have come in on one of Tuvalu's two cargo boats from the eight outer islands, a fairly rare gathering from across this widely scattered island country. Tuvaluan society has been picked apart, every aspect analysed, and plans are being developed, discussed in typical Polynesian consensus fashion, and slowly implemented.

The gardens in almost every yard on Funafuti are part of the NAPA, as are improvements in seawalls to protect the vulnerable atoll shores, steady improvements in freshwater collecting and storage, attempts to deal with the serious solid-waste problem depressingly obvious especially at both ends of the atoll's roads, testing of composting toilets as replacements for septic tanks, a re-visiting of solar energy as a supplement to ever more expensive diesel-generated electricity, reviewing the efficiency of the several government-owned corporations and informed and sensitive attention to the threat of HIV/AIDS in this very conservative Christian country. There are about eight or nine HIV-positive Tuvaluans, probably returning seafarers, but nobody talks much about who might be infected, though all are well aware of the threat the disease poses.

The effects of global warming here are not spectacular. They're creeping and insidious, weakening the already fragile fabric that enables this tiny atoll society to exist. Add more severe storms, or periodic – and entirely natural – higher-than-average tides, and the Tuvaluan environment could start to disintegrate. Insensitive human intrusions into the environment, and steadily more severe population pressure, don't help either. This is what NAPA aims to address.

Friday, February 24, 2006. It's raining this morning. But it's not a raging tropical storm that can hit Funafuti with sudden ferocity, a black wall of cloud rising from the sea and pouring thick rain on the atoll accompanied by deafening thunder and sharp cracks of spectacular lightning, and then pass as quickly as it came.

This morning's rain is gently soaking, the kind of rain many of the very religious Tuvaluans pray for. But much of this rain will turn into poison. Explaining why is a good way to explain some of the effects of global warming on this tiny country.

Tropical atolls often have a lens-shaped freshwater table just beneath the sandy, salty soil surface into which rain percolates, suspended above denser seawater that also seeps beneath the atoll. Many atoll societies drill into this lens and pump the water up. Local vegetation draws on this water as well.

In October 1942, World War II reached Funafuti and two outer islands when American forces occupied them, setting up port facilities and buildings, and blasting deep channels through reefs. On Funafuti Atoll, *pulaka* pits and hundreds of coconut and pandanus trees were dug up to make way for the first airstrip; rock and soil were taken from several large pits around the atoll. The "Seabees" assured locals they were only borrowing the material and the loan would be repaid.

These large pits are now filled with water and rubbish. They are called "borrow pits", or *taisala*, and because of population pressure and severe land scarcity, are now surrounded by houses or shacks. Some are even built over the pits. They are shattered parts of the atoll, and allow seawater to seep into the freshwater beneath.

More recent building, and steadily increasing septic-tank construction, increased the damage, and stresses, on the atoll. Add the stress from more and increasingly severe storms, surges from the open ocean and slightly but steadily rising sea levels, and you can see why Funafuti is a place where the rain turns into poison.

Freshwater security has been one of the major topics for discussion at this week's NAPA workshop – as are the problems *pulaka* farmers face trying to grow their large, slow-maturing equivalent of potatoes. The tubers are grown in pits a metre or more deep and the plants are carefully and expertly mulched and tended over two to three years until the large tubers can be harvested.

Now the *pulaka* pits, which you can see in a few places on Funafuti Atoll, with the tuber's distinctive, elephantine, dark green leaves, are suffering saltwater intrusion, causing yellowy tinges on the leaves, rotting the *pulaka* from beneath and stunting the growth of those plants that survive.

In the afternoon, I visit the Secretary to Government, Panapasi Nelesone, the country's most senior civil servant, the man responsible for Funafuti's preparations for next week's higher-than-average tide and its long-term economic and social survival. The country's Disaster Committee has already made preparations for evacuations from especially vulnerable houses, relocating evacuees in local *Maneapa* if the need arises.

While the predictions on paper suggest a peak of 3.26 metres at 5.30pm on Tuesday, February 28, with a new moon rising, actual projections based on January's peak, which exceeded the projections, and more recent sea-level data, are pointing to a peak closer to 3.4 metres. Nobody seems to be too worried.

Late morning, February 25. Saturday's a day for shopping, washing and cooking on Funafuti Atoll. The larger *fusi* in the central village of Fongafale is busy. Shoppers balance large bags of flour or rice on their motorbikes. Children are everywhere. Women are doing the washing, airing bedding on mats in the sunshine, clothes swinging from lines strung between coconut or pandanus palms in the almost non-existent breeze.

Slathered in 40+ sunscreen, with hat firmly planted on my head, I'm steadily regaining my sense of balance as I tool north up Funafuti Road on a small, single-gear motorbike that sounds like a large mosquito.

The breeze of movement is welcome; it's ferociously hot this morning. Nick, the husband of a local journalist friend of mine, and a mate, drive by and tell me they're heading to the end of the atoll to get some coconuts. I agree to meet them there later.

The atoll gets thinner the further north you go, so I see more of the Pacific Ocean on my right, through gaps between houses and *fales*, and *Te Namo* on my left, with houses, *fales*, and pig pens built precariously close to the usually calmer water.

Along the way, I see another friend, Ben, from the Government Information Technology Office – real dot tv, as opposed to the other websites with tv domain names leased from the Tuvaluan Government through the US e-commerce and security corporation, Verisign. This agreement helped pay for the paved roads such as the one I'm riding on this morning, and brings in about US$2 million ($2.7 million) a year in overseas income for Tuvalu's tiny, US$11 million economy.

Ben's wreathed in smoke from an earth oven, an *umo*, he's tending on the narrow lagoon shore just off the road, the pleasant smoke from smouldering coconut husks heating the white stones in the shallow pit wafting with him as he walks towards me.

"*Talofa*, Ben! What are you doing?"

"Killed a pig," he replies, smiling, pointing at the gutted and dressed pig draped over a low tree branch ready to be buried under the *umo*'s hot stones.

"It's our daughter's first birthday party tonight," he says. "Would you like to come?"

Children's first birthdays are celebrated with feasts and great joy here. Though medical care has improved enormously, not so long ago a child's first birthday was a good sign they would survive to adulthood. Ben's spontaneous invitation is typically Tuvaluan, and I gratefully accept.

He waves me off and I continue my journey north up Funafuti Road. Driving around Funafuti is like driving on Australian country roads. Everybody has a nod, a wave or a cheery smile as you pass.

Along the way, I see graves of Tuvaluans. Many are obviously very respectfully tended, pointing to the role ancestors play in Tuvaluan society, linking the living with their past in the central core of Tuvaluan existence, *Te Fenua*, the people, their culture, their island or even their particular part of their home island, past, present, with the children being their future.

Seeing Nick's car parked on the roadside near a thick plantation of very tall coconut trees, I pull over and wander into the knee-deep grass and welcome shade to find him carefully studying the palms to see which one is worth climbing today.

Behind me, over the narrow bitumen road, the relentless Pacific, with the tide just on the turn towards rising, is roiling against the atoll. The still blue-green *Te Namo* peeps through the trees and undergrowth in front. In the cool shade, a few mosquitoes seek out the unusual change in their Tuvaluan blood diet: me.

This area, and several others like it towards the ends of the atoll, serves as a local commons, where anybody can come and help themselves to coconuts, pandanus and leaves used for weaving mats, garlands, baskets and decorations when there's a *fatele*, a lively Tuvaluan celebration. The commons is respected, so nobody's greedy, and the area is thankfully free of rubbish.

Nick wraps some rope around his feet, hops up, plants his feet against the trunk of a palm and shimmies up. I move safely away to watch as he twists the green, head-sized nuts loose to fall with a loud thud below. With a dozen nuts on the ground, he shimmies down, unties his feet, husks one on a steel spike he's planted in the ground and hands it to me.

"*Fakafeti*," I reply, reaching for a pen from my shirt pocket to poke out one of the nut's eyes. Natural container carefully opened, I put the hole to my mouth, tip my head back and enjoy one of nature's true gifts.

Sunday morning, February 26. It's good manners on Funafuti Atoll for *palagi* to attend church, so I'm outside the Fongafale Church dressed in my sandals, *sulu*, white shirt and blue tie, nodding to folks wandering along the road to the biggest EKT church on the island.

Paying my respects to the pastor and his wife before the service, I discuss with them how locals are thinking about the looming high tide. He's not worried, saying God will protect the island and its people, and his wife nods in agreement. Many, especially older or more devout, Tuvaluans explicitly believe God's promise to Noah to never again flood the Earth, so talk of sea-level rise caused by global warming terminally threatening Tuvalu's very existence some time in the future is dismissed as *palagi* stuff.

The EKT, however, like all theologically literate Christians, also teach that bad stewardship of God's Earth is causing global warming and sea-level rises. At the NAPA workshop, another, younger EKT pastor attended, not just to open and close daily proceedings with prayer and say grace before lunch, but as a representative of a key Tuvaluan institution with a vital role in its survival. The consequences of bad stewardship of the Earth are close to home on Funafuti — with serious solid waste problems, rubbish dumps at both ends of the island, rusting vehicle hulks on the roadsides, garbage fouling the water-filled *taisala*.

The service is in Tuvaluan, and the sermon from Exodus. The pastor tells the worshippers how, in spite of everything, Moses steadfastly trusted in God to bring his people to the Promised Land. In the closing announcements, the pastor reminds his people that this week will bring very high tides and that the church will assist the disaster authorities, and asks for prayers that the weather remains calm.

While not disbelieving of the power of prayer, I also take science seriously, so later that calm Sunday afternoon, I braved the ferocious heat, walked across the blindingly bright airstrip, sandals crunching on the packed dry path to the Met office front door, and checked out the latest three-day forecast in the blessedly cool air-conditioned white bungalow: continuing fine, hot, light winds from the north-west or east, and a few passing showers – more of the same weather.

I walked back towards the airstrip and came across the weirdest sight I have ever seen. Only a few minutes before, I'd walked along a hot, dry, packed sandy track; now water was bubbling up through cracks in the track, spreading along the ground. I could hear the "blip, burble, pop" of the bubbling and seeping from the larger cracks.

The water was flowing across the track where I was standing in stunned amazement, deep enough to lap my sandals, my toes washed by water too hot to tolerate. The fearsome heat from the sun was all around me and now hot salty water was oozing as well.

I'd seen video and still pictures of this occurring outside the Met office before, and been told about it by witnesses, but to be standing in the middle of it was genuinely astonishing. I had to document this myself, so I ran back to my hotel room, grabbed my cameras, shouted for colleagues to join me, and ran back across the airstrip.

By the time we got back, the track was awash, the Met office's front yard rapidly filling, sheets of water forcing their way up and spreading south into the power station's compound, and north towards the Public Works Department depot. We waded in foot-deep, then ankle-deep, and then mid-shin-deep warm water to record the seepage.

Bemused locals, who'd seen all this before, were gathering on the airstrip to watch us and, yet again, see parts of their island inundated.

As arranged, a driver and a truck from the hotel found us and we jumped aboard and headed north towards a *taisala* and a nearby *maneapa* we'd heard was especially low-lying. Sure enough, water was lapping out of the flooded *taisala* along a side road and, across the pond, agitated pigs were grunting and squealing in their pens while their owners watched in case rescue was needed.

The Sunday tide was predicted to peak at 4pm at 3.1 metres and, as it rose, it oozed through the shattered atoll.

The calm weather, with no breeze pushing the waves, meant that the tide height would determine whether or not there was flooding as the lagoon overflowed its low bank. To the east, the ocean was kept at bay by the old berm tossed up by Cyclone Bebe in October 1972.

Along Funafuti Road, we glimpsed yards with water pooling in them, watched by locals worried their homes might be flooded, while life went on as normal elsewhere.

Along the narrowing road I'd travelled yesterday, *Te Namo* lapped dangerously against the shore, but it only broke through at one place. Houses planted on, and in some cases, even over the northern *taisala*, appeared to be safe, but a couple were in danger closer to the road.

On Sunday night, many Tuvaluans on Funafuti prayed that this was the worst they would experience over the next few days. At the Met office, Hilia Vavae and her staff, having waded to work through their flooded front yard, and studied the real-time data from the Australian-provided tidal monitor at the port complex to confirm the 3.1-metre prediction had been exceeded by a couple of centimetres, predicted Tuesday's peak tide would be even higher and that the weather would remain benign.

Tuesday, February 28. Bright yellow sunshine bathes the atoll, a gently cooling breeze caresses it and rustles the palm trees, and, with the tide approaching its lowest, the lagoon's colour ranges from bright green close to shore to deep blue. A few locals are lazing in the shallows or tending to their boats for a fishing trip later today. This glorious sight is tropical atoll paradise travel-brochure stuff.

But there was no glory for several households on the edge of a northern *taisala* last night, because, with a slightly higher tide peaking at over 3.2 metres just before 5pm, polluted water flooded into their homes, and they were evacuated by the Red Cross. Seawater seepage occurred around the atoll, but *Te Namo* and the Pacific did not break through to cause any serious damage. Some locals told how they'd caught fish washed into their kitchens from the overflowing *taisala*.

Today's the day, according to the predictions, that Tuvalu will experience the highest tide between 1990 and 2016, at 3.26 metres at about 5.30pm.

The Met office folks say the persistent high-altitude convergence over the country is making the weather benign, but some Tuvaluans no doubt believe *Te Atua* (The Almighty) heard their prayers and blessed them with good weather.

Funafuti's thin lifeline to the south, its thrice weekly Air Fiji flights, was severed again today. The flight was cancelled, this time due to vandalism at this end. Just south of the airstrip, the BBC crew that's here shooting segments for nature documentaries, has set up a time-lapse camera on a concrete slab near the entrance to a small cluster of houses and *fales* near the Assemblies of God church and hall.

Getting some "before" pictures, we head further south to where *Te Namo's* broken through near the end of the road. Only fifteen minutes have passed, but the little village is now deeply awash, with children running about pushing small pieces of wood through the flood like toy motor boats, watched by stoic adults hoping it doesn't rise any higher.

Further north, the road is awash and I force the car through the water in low gear, past adults up to their shins and playing kids. They've never seen it so high here, they tell me as I wade back.

Out along the long airstrip, the sports teams look like clusters of ants, occasionally tossing up splashing water as a player chases a ball to the flooded verges.

The Taiwanese garden supervisor is anxious about the seepage into his gardens. Earlier he soaked the place with fresh water to try to protect the plants.

More flooding scattered along Funafuti to the north. I'm reminded of storms back home, where some suburbs can be flooded or damaged, while nearby, evenings are entirely normal. Rather strange, driving north through 'suburban' Fongafale with its yellow street lights, towards flooded *taisala* and more scattered, local flooding seeping around some lower parts.

At the Met office, its front yard again awash in shin-deep water lapping at the front step, Hilia accesses the raw data from the tide monitor at the port complex, and exclaims that this has been a record high tide for Funafuti – just over 3.438 metres, exactly as she predicted.

We say goodnight and wade and then walk home, grateful to either the benign weather or God's benevolence, or both, that this high tide left Tuvalu and Funafuti relatively unscathed.

Friday afternoon, March 3. A late afternoon tropical storm is drenching Funafuti Atoll, pouring thick rain across the island from even thicker, darker clouds that blew in from the north-west.

The fresh rainwater will quickly disappear once the storm passes, topping up the water tanks, percolating into the polluted water table beneath the atoll and turning into poison beneath the ground.

The Prime Minister, Hon Maatia Toafa, and the Secretary to Government, Panapasi Nelesone, watch the storm with us and chat about the week and its record high tides at a time when international attention is increasingly focusing on global warming and the related rises in sea levels. They know that they cannot rely on prayer alone, and are committed to plans to ensure the survival of Tuvalu and its people.

The extremely high tides this week have nothing to do with global warming and sea-level rise, even though the seas have risen slightly over the years of detailed measurements. But the attention paid to global warming makes the world more acutely aware of the plight of those living in the vast Pacific Ocean.

Maybe the sustained active convergence zone that made the local weather so benign can partially be attributed to global warming, but the models used are just not sensitive enough to be useful even over a large area like Tuvalu's 900,000 square kilometre exclusive economic zone.

If the extreme tides had coincided with a storm like the one that ended the week, there would have been much worse flooding and more damage. Global warming and its predicted effects on Tuvalu will weaken an already fragile environment vulnerable to very high tides, storms and sea surges, with damage amplified by erosion, groundwater pollution and seepage, loss of vegetation and weakened reefs. Tuvaluans with sympathetic *palagi* assistance are not waiting for this to happen. Instead they are struggling to ameliorate the worst effects and develop a long-term survival plan, improving the health and education of the children and providing them with the means to draw on the best the world has to offer. ■

Mark Hayes is a Brisbane-based journalist and journalism educator.

Essay:

Seven-tenths: Random notes from the deep

Author:

Creed O'Hanlon

Image: John Lewin, 1770–1819 / Fish catch and Dawes Point, Sydney Harbour, c.1813, oil on canvas / Courtesy of Art Gallery of South Australia

Seven-tenths: Random notes from the deep

One way to open your eyes is to ask yourself, 'What if I had never seen this before? What if I knew I would never see it again?'
– Rachel Carson, author and ecologist

The second time I crossed an ocean under sail, I had first to fly across it. I was one of a small crew of three bound for Fort Lauderdale, on the south-east coast of Florida, from where we were to deliver a forty-five-foot timber ketch to Malaga in Spain. It was 1980. The owner, an Englishman who had made his money dealing in second-hand aircraft, had secured our services just six weeks before the beginning of the Caribbean hurricane season. He had booked us on the cheapest flights he could find: an Air India service to New York from London's Gatwick Airport, connecting with some bankrupt, no-name shuttle service to Miami. The initial flight followed a great circle route looping over the Arctic rim of the North Atlantic. The early spring skies were clear. I had six hours to contemplate what a voyage across all that empty water would be like.

From 10,000 metres, only faint specks of white, the crests of the largest, wind-blown waves, glinted on the pale grey-blue expanse. When a large ship was sighted, every few hundred kilometres, it was like a tiny, russet insect. Nothing smaller – neither an iceberg nor, especially, a yacht – was visible. It was hard to imagine what living just a few feet above the ocean's surface, far from land, for nearly a month would be like. I knew only in theory that survival would depend on a precarious, imperfect conjunction of weather, seamanship, navigation, endurance, and a watertight hull.

All I could think about was the deep, the kilometres of cold dark water that clawed at a vessel's keel from the ocean's bottom: to me, the thought of this was more disturbing than the intractable vastness of its surface, whatever the state of its swell, the speed of its shifting tidal streams and currents, and the unfettered strength of its winds.

A week later, we sailed beyond the edge of the North American continental shelf. When the line of soundings plunged from less than a couple of hundred metres to a couple of thousand, and the last smudge of low-lying land slid beneath the horizon. I had to stifle a sudden, atavistic fear of the open sea by reducing id daunting reality to statistics. The few tonnes of salt water our vessel displaced – right there, on a western eddy of the Gulf Stream, where tentative 'x' in soft pencil 'fixed' the position of our departure east-bound across the Atlantic – were a minute fraction of the 1.37 billion cubic kilometres that covered roughly seventy-one per cent of the planet, an area of around 361 million square kilometres. The numbers were as abstract and as barely imaginable as the British Admiralty metric charts on which land was always a flat, bilious yellow, inshore waters an insipid blue, and seas beyond the 200 metres line of soundings white.

Except in those moments when we are immersed in it, or floating on it, the sea's expanse is almost incomprehensible. We are awed by its power and limitless mutability, and we ascribe to it aspects of human mood and even sentience, usually during those episodes when it rises to inflict its force – note here the ready use of an emotive verb inferring mindfulness, even cruel intention – on us well above the high water mark that is the nominal demilitarised zone between a marine environment and 'dry' land. No matter how much we love the sea – or claim a near-mystical empathy with its chimeric mammals, the whale and dolphin – very few of us feel for its alien and uncompliant ecology the same intense intimacy, that visceral sense of connectedness, of elemental dependence, that we do for the landscapes of our natural habitats ashore. Maybe it has something to do with uncertainty, a fear not so much of the unknown as the unknowable.

Much of the sea is invisible to us. Its most spectacular topography is unreachable, lying at depths well beyond the capacity of humans to reach without expensive and cumbersome mechanical support. Tens of millions of acres of submarine flora close to well-populated shores on every continent are as unknown and undocumented as the tens of thousands of kilometres of vertiginous oceanic trenches and mountain ranges that surpass the Himalayas in scale. Apart from a few functional structures such as piers and breakwaters, offshore lighthouses, and oil rigs, humans have failed to impose a permanent architecture on the open sea – there are none of the boulevards, plazas, parks, cathedrals, castles, and monuments that impress upon us the ingenuity and creative accomplishment of a large city. The sea is untenable. Our place in it is always fleeting and artificial.

Which only increases our fascination with the fish, mammals, reptiles, crustaceans, jellyfish, starfish, corals and anemones that inhabit it, even if the reality of our relationship with them is defined less by curiosity or concern than by a ruthless, industrialised harvesting that has turned us into the ocean's most voracious predators.

Again, whatever uneasiness we harbour about this is assuaged by the oceans' unrestrainable spill. From the air, where most landlubbers get their first view of an ocean unbounded by land, it appears too big and indomitable to be despoiled by mere human activity. Looking seaward from a heavily urbanised stretch of coast like Long Beach, California, or Yokohama, Japan, or closer to home, Sydney – beneath orange-tinted skies laden with dust, smoke, and chemical emission – the shimmering surface appears pristine and undisturbed by everything but the wind.

A quarter of a century before statistical reports on the state of the marine environment began to pile up like past due bills on the desks of politicians, journalists and academics, there was already plenty of anecdotal evidence of the oceans' deterioration. For half a dozen years, during the late 1970s and early 1980s, I worked as a professional seaman on vessels on both sides of the North Atlantic Ocean, on the Baltic, North, Mediterranean and Caribbean Seas, and along the Pacific coasts of the United States and Mexico. Back then, dockside mutterings about the lack of sighting of larger pelagic fish such as tiger sharks or marlin, or of populous migrating pods of whales and dolphin, or of flying fish, which every day used to collide, mid-air, with the pitching decks of ocean-crossing yachts (and end up in the galley, being prepared for breakfast), were common. Yachtsmen and the crews of inshore fishing trawlers also worried about being sunk by half-submerged steel containers – thousands are still lost overboard every year by merchant ships during bad weather – or disabled by plastic bags sucked into engine intakes or polypropylene ropes snarled around propellers.

Occasionally, some of us would bear witness to a small catastrophe. In 1977, I was on watch aboard a sea-going tug as it approached the Bay of Naples at the end of a two-day passage from Malaga in Spain. A dense pod of beak-nosed common dolphin surfaced about half a kilometre off the starboard bow. In a frenetic sequence of arcing leaps and splashdowns that churned the glassy swell into white water, they led us between the islands of Ischia and Capri towards the Italian mainland. A few nautical miles from Naples itself, the blue water turned brown. A dolphin swimming ahead of the rest of the pod, as if taking point, a solitary scout, leapt high out of the water and with a sickening, high-pitched squeal, fell sideways into it again with a graceless splash. It bobbed lifelessly in the low swell. Several more dolphins began to thrash on the surface nearby, then, one by one, they rolled onto their backs and lay still. Suddenly, the rest of the pod swerved away and sped seaward, no longer leaping, intent only on finding safety. A couple of minutes later, the tug sailed past a few dozen of the dead, which wallowed in a greasy slick of foul-smelling chemical a couple of hundred metres in diameter.

We had been out of sight of land for nearly a week, sailing north-nor-east towards Bermuda with the full strength of the Gulf Stream flowing with us, when I realised how dark the ocean was, even under clear, moonlit skies. The usual nocturnal bioluminescence, the phosphorescent shimmer of microscopic organisms such as plankton floating just below the surface, appeared to have faded. Without it, the water beneath us felt more desolate and unwelcoming.

During the day, the ocean, now the dull blue-black of a fresh bruise, was just as empty. A pair of humpback whales, a mother and calf, breached close off our beam as we skirted a reef protecting the harbour of St George, in Bermuda, and a familial pod of dolphins gambolled in our bow-wave for a few hours just west of the Azores, but we sighted nothing else during the month we were at sea. The daily routines of voyaging were punctuated by a disquieting frustration that the ocean was emptier than imagined it would be.

Halfway across the Atlantic, we found ourselves hove-to for nearly two days in a strong south-easterly gale, a full Force 9 on the Beaufort scale, in an area where *Ocean Passages of the World*, the staple reference work for professional navigators published by the United Kingdom Hydrographic Office, insisted gales from the south-east and east were virtually unheard of. The Azores High, a large, permanent high pressure system centred over the Portuguese-controlled islands nearly a thousand nautical miles west of the Portuguese mainland, had slipped unexpectedly far to the south and the isobars of its north-west quadrant were constricted by a wave of deep low pressure systems moving north-east towards the English Channel. The storm eased just before it forced us to surrender our track towards the Azores port of Horta and make, instead, with rationed food and water, for Coruña, on the north-west coast of Spain, another week's sailing further on.

The increasing instability and intensity of seasonal weather systems, as well as localised fluctuations in ocean temperatures and atmospheric pressure, threaten those who make their living on the sea – and the thousands of amateurs who set out on offshore passages in small yachts (aided by satellite global positioning systems that obviate the need for old-fashioned skills and old-fashioned tools such as a sextant and a chronometer) – with ever more violent, unpredictable weather.

During last year's West Atlantic hurricane season, the United States' National Oceanic and Atmospheric Administration named twenty-seven tropical storms, among them fifteen hurricanes. Three of the hurricanes were devastating Category Five storms, and at least three of the tropical storms misbehaved in ways that were unsettling to everyone who thought they understood the weather systems in the North Atlantic's sub-tropical latitudes. Tropical storms Epsilon and Zeta sprung up outside the traditional hurricane season, the latter on the second last day of the year, exactly a month after the season's official end. A late season tropical storm, Delta, wandered eastward across the Atlantic to batter the Canary Islands, off the coast of Morocco – a 2,000 nautical mile detour from the usual track of such depressions north along the eastern seaboard of the United States or across the Florida panhandle. Those who sail in higher latitudes have observed that low-pressure systems there are deepening and the winds are blowing harder.

The elemental behaviour of the sea itself is also changing. The fast-moving Gulf Stream that carries warm water from the Caribbean across the North Atlantic – causing winter temperatures in Reykjavik, Iceland, to be a little higher than in New York, and tempering the worst effects of high latitude depressions that track across Ireland and the British Isles throughout the year – is beginning to slow. The mooted long-term effect of this – Northern Europe might be beset by a 'big freeze' that could persist for several centuries – will be aggravated if the Arctic's sea ice continues to recede and the slow melt of Greenland's giant glaciers persists or, worse, accelerates.

Fifty-five years ago, the late Rachel Carson, one of America's first environmental activists, assembled a series of her writings on the nature of the sea in a best-selling book, *The Sea Around Us*. Its tone was somewhat overripe by today's standard, and its scientific facts have become, inevitably, amusing anachronisms, but the author's confidence that the sheer immensity of the sea might protect it from the worst that humans would inflict on it was, for an American postwar generation steeped in the social idealism of Truman's Fair Deal, and an understandable desire for renewal, compelling. "For the sea as a whole, the alternation of day and night, the passage of the seasons, the procession of the years, are lost in its vastness, obliterated in its own changeless eternity," she wrote. Her confidence was inspired by a more innocent time. Today, we are numb to the stark symptoms of stress and deterioration that are apparent in nearly every ecosystem. We pay only lip service to doing what it takes to alleviate them.

There is, in the florid closing lines of Rachel Carson's book, an unintended glimpse of a grimmer scenario if we do not try harder to turn things around: "... [the sea] encompasses all the dim origins of life and receives in the end, after, it may be, many transmutations, the dead husks of that same life. For all at last return to the sea – to Oceanus, the ocean river, the ever-flowing stream of time, the beginning and the end." ■

Creed O'Hanlon writes regularly for *Griffith REVIEW*. His most recent essay, 'Mixed blessings' was published in *Griffith REVIEW 10: Family Politics*.

Essay:
Corals under siege
Author:
Rosaleen Love

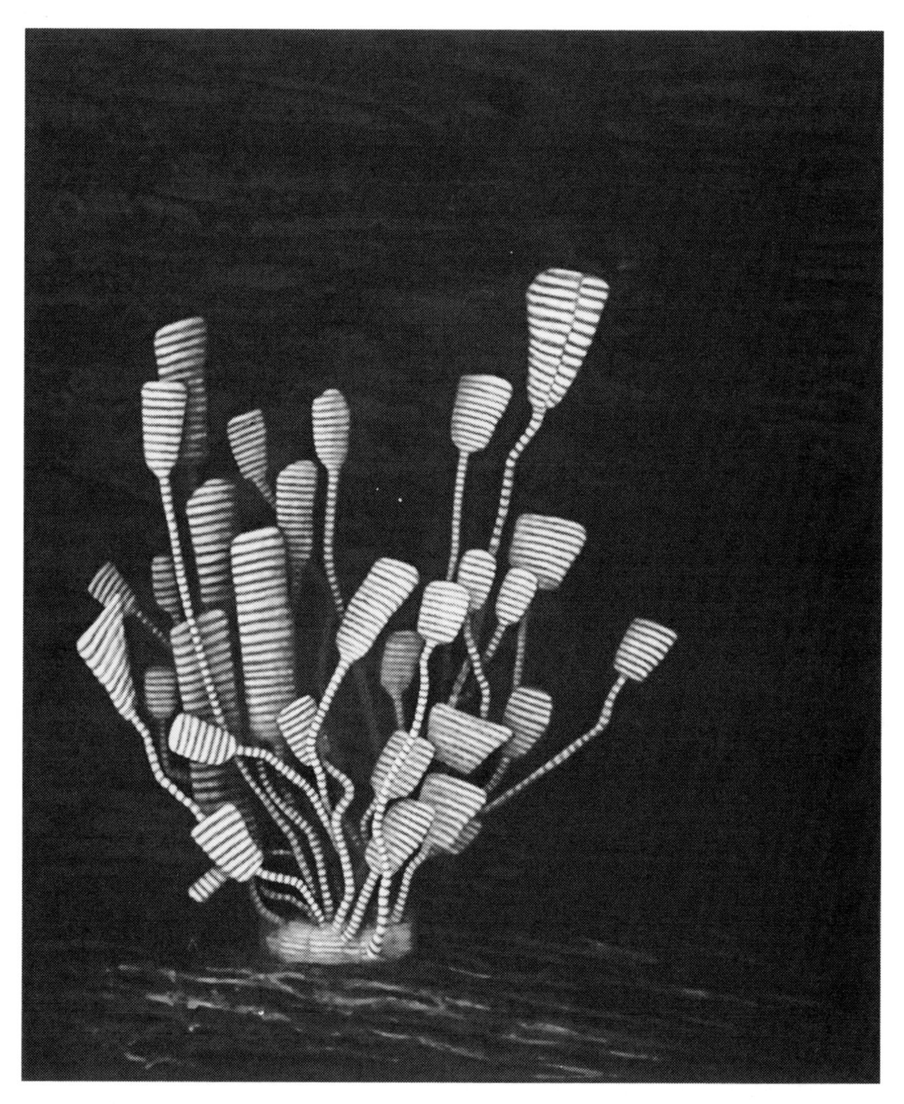

Image: Vera Möller / Hummington 2005 / Oil on canvas 92 x 76cm / Courtesy of the artist

Corals under siege

There's not much of a laugh to be had on the topic of global warming but American futurist Bruce Sterling does his best. Sterling's weapon is satire; his tools include the blog and the after-dinner speech. In his "Viridian Manifesto" (*Whole Earth*, Summer 1999), Sterling calls on artists and designers to join in resistance to the forces of global warming. Viridian, for Sterling, is both an aesthetically pleasing shade of green, and also the colour of Big Mike, the Viridian mascot, a micro-organism that, in death, decays and does the elegant recycle thing. The Viridian movement is a niche inhabitant of the wider green scene, one devoted to taking the mickey out of the pronouncements of the military–industrial establishment while promoting good design as one useful response to the global warming challenge.

While they may not yet have brought the establishment to its knees, contributors to the Viridian website exhibit a ferocious lateral thinking that is refreshing in these post-Montreal days of political posturing while Rome burns. The Viridian movement, Sterling says, is all about "creating irresistible demand for a global atmosphere upgrade". A quixotic enterprise, this task may well be doomed (Sterling is no optimist), but it foregrounds the flip side to all technology upgrades: their wider impact on the atmosphere which, by its nature, is a global entity.

As the warming atmosphere interacts with the warming ocean, both affect the health of coral reefs. It is becoming increasingly clear that only a global atmosphere upgrade is going to save the Great Barrier Reef as we know it and love it.

The news is not good. Here are some recent headlines from news reports: "Great Barrier Reef to be decimated by 2050"; "Climate change: icons under threat. The Great Barrier Reef"; "Coral bleaching, the reef's biggest threat"; "Too late to save the reef". *Science* reports a glimmer of hope with the bouncy word "resilience" ("Climate change, human impacts and the resilience of coral reefs"), but the real issue is how much resilience will be enough. These dire warnings about global warming and the Great Barrier Reef are read in the context of equally alarming climate news: melting glaciers from the Arctic to Bhutan; a predicted rise in the rate of species extinction; dramatic changes in the strength of the currents that drive the Gulf Stream, and more.

What is becoming apparent is that the past 10,000 years of human history may well have been an unusual period of climate stability. Climate instability may be the norm and we are the people who may be lucky, or unlucky, enough to be both implicated in the coming changes and affected by them.

Long before life took off on the land, there were reefs in the oceans. Reefs were different then. Four hundred and fifty million years ago, the reef builders were not the corals of today but sponges (the stromatoporoids), sea mosses (the bryozoans) and other classes of organisms, many now extinct. The earlier forms of coral, the tabulate and rugose corals, were like modern corals in that they formed hard reef-constructing skeletons that provided niches for other forms of reef life, like soft-bodied creatures that left no fossil record. Remnants of these ancient reefs can be found, now well above the sea, in the Canning Desert and the Gogo region of north-western Australia or, closer to the present Great Barrier Reef, there are the fossil reefs of Charters Towers in North Queensland.

Finding a reef high and dry and stranded in outback Australia brings powerfully to mind the instability of landscape, the shifts back and forth from former seas to present land, from land now to future seas. Long before the human era, the Gogo reefs grew and died and left their structures in the shape of inland atolls, their relics in the fossil fishes that lie scattered on the surface of the plains. Earth's time scale is here read in the landscape of fossil reefs. Life evolved in the oceans, and reefs existed then in full, if different, complexity. Ice ages came and went, continents collided, reefs eroded when exposed to air in rock uplift and then grew again as ocean floors subsided. Reefs coped with change and, in turn, helped create change, change to ocean currents and chemistry – even, perhaps, changes to climate.

Something new is entering the story. Reefs are changing again, and now it is the turn of humans, rather than fish, to bear witness. What is new is the effect that human activities are having on reefs. While overfishing, pollution and plagues of the crown-of-thorns starfish are serious enough in themselves, ocean warming is proving worse. The year 1998 was the hottest year on record for a thousand years (until 2005) and the summer of 1997–98 was also the year of the stress and death of corals through coral bleaching on a scale never seen before. Early in 2002, a further outbreak occurred, more serious than in 1998. The widespread death of corals occurred at a level called unprecedented, not only in the history of human life on earth, but also, in its suddenness, in the fossil record.

The principal cause of coral bleaching is global warming. In bleaching, coral loses its colour and much more. The brilliant colours of corals come from the symbiotic relationship that is the coral polyp, where one partner, the plant, lives inside the other, the coral. The plants are tiny single-celled algae, the zooxanthellae or symbiotic dinoflagellates that live within the tissues of corals in great numbers. When the zooxanthellae become stressed by the heat of the oceans, they collect in the hollow column of the coral polyp and leave their host. The coral skeleton becomes visible through the transparent polyp and large areas of coral turn white. As corals lose their colour, they lose some 60 per cent of the energy derived from the symbiotic relationship. When the zooxanthellae leave, ill-health and often death of corals follows. Huge areas of reefs worldwide whiten and frequently die. In the nature documentary *Silent Sentinels*, the sight of huge areas of dead white coral makes for compelling viewing of the disaster-movie kind.

Coral bleaching is relatively new as a cause of massive death of corals. In 1998, reef scientists were unsure about what would happen next. Was the massive bleaching a one-in-a-thousand-year event or was it a sign of more serious damage to come? In 2001, the Intergovernmental Panel on Climate Change pronounced that global warming is a reality, and as a result, ecosystems will change. Within a generation, reefs are expected to change dramatically. Not all corals will die, and some will recover and adapt, but it is the rapidly increasing frequency of global bleaching events that is of great concern.

In all the previous talk about greenhouse warming and extreme weather events, this particular event was not predicted. Earlier, in 1987–88, the CSIRO and the Commission for the Future organised a huge public awareness campaign about the impacts of the human-enhanced greenhouse effect on Australia. The impact on coral reefs was part of the brief, but more in the context of rising sea levels, and the impact of ocean warming was considered minimal. What science did not predict, and could not have predicted given the knowledge in 1987, has emerged as its most dramatic challenge. In addition, if the amount of carbon dioxide in the atmosphere doubles between 1990 and 2070, as may happen, it will also increase in concentration in the oceans, and this in turn will affect the rates at which corals take up calcium for their skeletons. The result is a form of coral osteoporosis, with reduced coral growth and strength.

The Great Barrier Reef has many meanings, with science providing one set of creation–destruction narratives. In indigenous stories, sea places have mythic stories to tell of how this place came into being, of how other worlds connect to this world. The stories of coastal Aboriginal people are tales of sea creatures and their journeys, stories that connect past mythic events with present coastal land and reefscapes. Here a spirit ancestor chased a whale or dugong, there it lay down to rest, and if you look with the eyes of faith you can see its shape in the rock and its breath in the spray of the waves – the reef shimmers with mythic significance.

In European times, the reef has been an economic resource for commodities, tourism and extractive industries. The reef has seen shelling and sheep farming, whale and dugong hunting, pearl and *bêche-de-mer* fisheries, penal and leper colonies. Limestone and corals were mined, people and drugs smuggled and oil drilling was only narrowly avoided. Events of the relatively recent past merge into the present growth and, some might argue, incipient decline of million-dollar tourist industries, where the reef gains new meanings as a place of pleasure and leisure, and the new jobs of reef managers and conservers are created.

Stories overlap and intertwine. On one level there is the litany of problems. For example, here are the problems a coral reef faces: crown-of-thorns starfish, agricultural run-off, and over-fishing. The reader is bombarded with issues and the recital of quantitative information and statistics, most of it gloomy. At another level of story comes the recital of cause and effect. If these are the issues, what are the causes? What chemical and biological factors, such as global warming, contribute to the destruction? The third level includes theories of biological and geological evolution. How has this reef system come into existence and how has it changed given the vast time scales of biological and geological epochs? The fourth level concerns the myths and metaphors of the reef, talk of the balance of nature, the resilience of coral reefs, the harmony of nature, the integrity of an ecological system (imagine also imbalance, discord, lack of integrity). How relevant are concepts of human morality, probity and prudence when applied to non-human systems? Reefs symbolise, on a metaphoric level, the creative–destructive nature of the universe. All these elements enter into ways of knowing the reef, ways of talking about it.

The attempt to come to grips with the issue of global warming and the Great Barrier Reef soon moves beyond the science and politics to their personal meanings. Why do we consider the Great Barrier Reef to be a beautiful place? If biodiversity is lost, what else goes? The term "biodiversity" is a highly political word, embracing not only the complexity of ecosystems but also their beauty and the moral dimension of how humans should act in preserving species diversity.

In the coral reef, biodiversity is made visible to the visitor: the underwater world seems crammed with forms of weird and wonderful life. The observer is highly privileged. Fish may keep a certain distance but do not flee from human encounter in the way that wild animals do on land. On a coral reef, biodiversity packs a profound emotional punch.

The beauty and intricacy of reefscape, of underwater landscape, is something that draws people into it, enthralls them in the tourism experience and has them coming back for more. Nature tourism may be given an economic value but there is more to it than this; many reef visitors find a quasi-spiritual meaning in the experience. The multiple character of the reef experience from science to tourism is an important factor in this discussion. Coral reefs also have aesthetic and emotional value. Reefs are ecological systems important for human health and wellbeing in more than the economic sense.

Historically, coral reefs were fraught places for mariners. When Captain James Cook sailed through the Great Barrier Reef in 1770, he did not, as far as we know, ever take mask and snorkel and have a look at the underwater landscape over which he sailed. He used clues from the surface to estimate the depths. He was understandably keen to put as much draught as possible between his keel and the dangerous reefs below. What nature revealed to him, through its surface, was: "Danger. Take care."

In the history of art, landscape painting has a long tradition. Unsurprisingly, there is no similar tradition of painting underwater landscape, as the zoologist William Saville-Kent pointed out in his book, *The Great Barrier Reef of Australia* (1893). Saville-Kent pioneered the art of reef photography, the photographer being better able than the artist to take advantage of the short periods of time the reef was exposed at extreme low tides. He captured reefscape as if it were landscape, with corals, clams, sea urchins and lagoons as elements in composition, in place of the peaks, rivers and forests of landscape art. His reefscapes lie exposed in air. As the first Commissioner of Fisheries for the colony of Queensland, Saville-Kent argued that the resources of the Great Barrier Reef should be harvested within conservation limits – he was an advocate for sustainability and biodiversity long before the words were coined. Saville-Kent's images helped create one concept of the beauty of coral reefs – the coral garden, with decorative hard corals predominating.

In the 1960s, the poet and conservationist Judith Wright was active in the protest against plans by the Queensland Government to permit oil drilling and limestone mining on Ellison Reef. In *The Coral Battleground* (Nelson Australia, 1977) Wright explains: "Slowly but surely as the years go on, we are destroying those great 'water-gardens', lovely indeed as cherry-boughs in flower under their once-clear sea ..." Cherry boughs? Water gardens? The mystery of the language is revealed in its cited source, *Five Visions of Captain Cook*, the poem written in 1931 by Kenneth Slessor. His poetic invocation of the "crystal twig", "petal", "water-garden" and "cherry-bough", springs from a European Arcadian tradition ill-equipped to cope with a radically different underwater landscape.

Judith Wright's passion to save the reef sprang more from her conservation principles than from personal experience. In the days before mass tourism, it wasn't so easy to visit the reef. Wright visited only once, spending a few weeks on Lady Elliot Island, a coral cay where the spectacular underwater landscape was in stark contrast to the land above water, an island ravaged by guano mining and goats. In the early years of reef tourism, the reef was explored from above, as it were, at low tide when reef walkers turned over lumps of coral rock to view the life beneath as it lay exposed in air or in shallow pools. The total immersion experience with mask, snorkel or scuba was not so common. Wright paid tribute to Slessor, but she took her meaning further. Her "water-gardens" were "far more complex, far more alive, teeming with myriads of varied animal lives". She moved from his early twentieth-century conceptual understanding of landscape-as-scenery to the science-informed landscape-as-environment.

Imagining reefscape in terms of underwater environment conveys both a sense of ambience in terms of sensory immersion, and also a sense of natural relationships in which the viewer is also a participant. The viewer may be there, underwater, with snorkel or scuba or may be an armchair traveller watching a nature documentary. For the diver, immersed in another world, water flows over skin, light ripples down from the surface through the refracting air-water barrier; the sound of one's own breathing is magnified; bubbles of air plop past. The myriad of varied animal lives continues, indifferent to the diver's presence.

For the armchair viewer, a sense of "being there" is heightened in hyperreality. The Australian documentary *Coral Sea Dreaming* (1992) brings the hidden sex lives of coral polyps to the screen. The viewer goes on an exhilarating ride with the coral gametes on the one day in the year when they are released in a mass coral spawning. As the larvae then rise like golden globules in the dark blue sea, we follow them to the surface. Suddenly we are above the water, looking down on the sea. The coral larvae look like a vast spread of brown scum, swirls of dishwater froth. To the next cut, where the view is from high in the sky. What was once scum becomes visible as the pattern of reefs in the sea. The film editor turns beauty to ugliness, ugliness to beauty in deliberate challenge to our aesthetic prejudices.

On the Reef Futures website (*www.reeffutures.org/topics/bleach/reefstate.cfm*), there is a coral bleaching simulator, where the visitor may enter the variables and see what the reef will be like in the year 2050. The website lets us choose from a range of conditions and presents us with images of the reef as it changes from being the present day "hard coral dominated reef" to a reef where the hard corals are in decline, and soft corals and algae provide more of the cover – a do-it-yourself ecological disaster simulator. The fear is that the Great Barrier Reef will become increasingly like Caribbean reefs, where hard corals have largely disappeared.

In December 2000, I travelled to the Lihou Reefs, a reef system outside the Great Barrier Reef, a remote, rarely visited part of Australia about 600 kilometres east-south-east of Cairns. It is the largest reef structure in the Coral Sea. In 1982, the Lihou Reef National Nature Reserve was declared a category 1a strict nature reserve, to be preserved in as undisturbed a state as possible. Cited in support was the aesthetic value of the underwater landscape: the reef system has "spectacular and unusual underwater topography and reef structures".

At Nellie Cay, we dived down the side of a coral bommie to the sandy reef floor, then drifted up past a steep wall of sponges, soft brown-green in colour, cups stretched towards the surface. Black feather-stars and colonies of lime-green encrusting corals contributed to a soft and muted reefscape, rather like Ireland in the mist, were it not for the fusiliers, damselfish and Maori wrasse that swam around. The reef wall with its muted blacks, browns and greens was a far remove from the coral reefscape as conventionally depicted, for example, by William Saville-Kent. In reef images the hard corals are selected for impact, often the *Acropora* species with their pointed twig-like fingers. The Lihou experience was an experience of a different, softer reefscape.

The Lihou Reefs also support large populations of soft corals. What on the Great Barrier Reef is often a sign of reef degradation must, in this remote place, be part of a more natural, but different reefscape. When a reef is damaged, whether by storm or human impact, soft corals are often the first species to recolonise, and their initial success may crowd out hard corals that previously flourished. The reef, after damage, will change its proportions of reef communities. At the Lihou Reefs, it is more likely to be the way things are. Yet a reefscape of soft corals has its own aesthetic value. A tourist may well take delight in them for their own sake, entranced by the colours and unaware of the links with cycles of destruction and creation. For the scientist, however, knowledge and appreciation of the underwater landscape are intertwined. Knowledge of loss privileges what is judged more pleasing, more pristine.

The reef management plan seeks to preserve the Lihou Reef system in its "near pristine" condition. Earlier in 2000, there was some talk of oil exploration in the area, as the Lihou Reefs are outside the Great Barrier Reef Marine Park, but such talk seems to have stopped, at least for the moment.

The science in management plans, such as the plan for the Lihou Reefs, provides a new frame for understanding and seeing beneath the surface. Appreciating underwater landscape involves an element of "seeing" the depth of time. The Lihou Reefs are special in part because two separate episodes of reef building created the present structures, but this will only be visible to the geologist who can see in present structures the outcomes of past events and processes. In the nature documentary, the narrator helps spell out these connections and tells a story of the dramatic progression of events and species through time.

The reef provides the vision of beauty; reef science supplies the theoretical links. Divers can see for themselves the interdependencies once they know what to look for. The anemone fish shelters within the waving fronds of the anemone host, enticing other small fishes into the anemone's trap. I see utterly different creatures – prickly sea urchins, sausage-like *bêche-de-me*r, sea stars, brittle-stars and feather-stars – and find science has discovered their close relationship within the phylum of the Echinoderms. I know that all the crannies in the corals, within corals, underneath coral rubble, are occupied by forms of life, because the photographer-naturalist has shown me in advance. Fishes glide languidly by and their indifference to human presence provides our pleasure.

Reef environments evolve over time, created by the activities of coral polyps and by forces of wind and water, ice and global warming. This is the "deep time" of the evolution of the planet, the oceans, the life forms that create and inhabit the reef. When Aboriginal people first came to Australia, perhaps some 40,000 years ago, the Great Barrier Reef as it is today did not exist. Instead, a coastal plain 200 kilometres wide extended to a limestone ridge on the shore, that ridge itself a relic of precious barrier reefs. When Judith Wright described "the battle to save that thousand-mile stretch of incomparable beauty from the real destroyers – who are ourselves", she was only partly right. Human destruction is real enough but it is nothing in comparison to the great reef-destroying episodes of the past, well before the people came.

The sad fact is that, as knowledge of global warming has increased and as ecological indicators are getting worse, human behaviour has not changed. There will be climate winners and climate losers, and who will win and who will lose is not yet clear. Reefs come and go in great cycles of time, grander than mere human time. They will be here, in some form, when humans are long at rest with the dinosaurs. But though change to the reef may well prove insignificant in geological time. I live in human time. I want the reef to survive in my time and for the human generations to come.

How can we put some kind of collective value on the individual experiences of so many reef visitors, how can we make the leap from the subjective experience of delight to the social dimension of conservation? Experience brings something, but it is not enough. Science brings knowledge but with the proviso that what there is to know always exceeds what is known, and probably, ultimately, the human capacity to know it all. The delight in beauty, the imperfection of knowledge and the desire to preserve the reef – here desire encounters conservation policy in the gap between what people say as individuals and what people collectively do.

Conservationists are concerned that their listing of what's wrong with the planet, the litany of environmental doom, may cause people to switch off from their message. Conservation talk may inspire, unintentionally, a feeling of despair in the listener, rather than a positive response to the struggle. The conservation perspective is one of increasing pessimism about the survival of diversity in species and in ecosystems. Ecological awareness tends to bring about a general gloominess of outlook. It's hard to remain happy knowing that the Great Barrier Reef may not be there in fifty to a hundred years.

Delight in underwater landscape has the power to lift the viewer from despair to active contemplation. On coral reefs, the ocean-air systems, photosynthesis, food chains, genetic codes, reproduction, speciation and its necessary companion, extinction, were in place long before humans arrived. The sea arouses the sense of immersion in ongoing life, immersion in a non-human frame of reference, and with this comes a sense of liberation. I am suspended in warm water as if I were a part of it, entering into some pre-human condition of flowingness. The history of human evolution rolls back for this moment of re-entry into the oceanic past, and the planktonic ego is liberated, for the moment, free from responsibility for the earth. Air and sky are left behind, the sea washes over, and it is a different self that dives down to meet whatever moves below. ■

A version of this essay with footnotes and references is available at www.griffith.edu.au/griffithreview

Rosaleen Love is the author of *Reefscape: Reflections on the Great Barrier Reef* (Joseph Henry Press, Washington DC, 2000). Her essay 'My friend the fridge' was published in *Griffith REVIEW 10: Family Politics*, Summer 2005–2006. She holds senior research associate positions at Monash and Latrobe Universities, Melbourne.

Fiction:
Flame bugs on the Sixth Island
Author:
Patrick Holland

Image: Neal Joseph / Rock Pool Memories / Oil on Canvas 80 x 115cm Courtesy of the artist.

Patrick Holland

Flame bugs on the Sixth Island

Go down to the rock pools when the evening tide is out and there is a chance you will see them. Sometimes one will swim in among the mangroves in the tidal flats but the rock pools are best. Flame bugs are what we call them. I do not know if they have other names. I do not know where else they are found but our island. I have never heard them spoken of by anyone who doesn't live here.

The north-east wind comes in spring and blows the flame bugs to our shores. One October in boyhood, I took to going down to the rocks alone to look for them.

I never asked the boys to come with me. If I had asked one, the others would have been jealous, not for any particular fondness of me but only at being left out. Also, I was worried they would try to catch and torment the creatures, though they are rarely caught by hand. I thought about how those boys dragged mud crabs out from under rocks with hooks and tried to crack their shells.

The most precious time I went looking for the flame bugs was with the girl we called Shell. We called her Shell as before anyone knew her she was seen collecting shells on the south beach, and because she wore a necklace with a by-the-wind sailor pendant. She belonged to that tribe of children whose European blood naturalises here; whose blonde hair the sun and saltwater turns white and the white skin olive.

One afternoon I saw Shell sitting bored in her front yard and, though I had planned to go alone, I asked her if she wanted to come look for flame bugs and she said that she did.

We left Ooncooncoo Street at twelve years old and six o'clock.

Shell had only recently moved to Moreton Bay, which is so close yet so far apart from the big city. She was lonely and a little intimidated at school. Most of us had grown up together and there were more boys than girls and we boys were very rough unless isolated. First I pitied her. Then I wondered at her: at her way of sitting with her knees beside her; at her speech and her interests that were cultivated and strange to me. I made a habit of noticing her. But I did not know how to introduce myself. This night looking for flame bugs was the first time we had truly spoken. Walking off her street I got the feeling she was excited at the prospect of making a friend, even of me, and that she would have followed me anywhere; far further than the rock pools.

She told me how at her old school she had played the violin but here there was no teacher. Her mother was doing her best in a proper teacher's stead. She told me she liked the island but for that. I told her I knew a girl who played piano, which was true. I told her my mother, being a school teacher, could let us into the community dance hall any time we liked, where there was an assortment of old instruments and the opportunity to nurture a band, which was not true at all. Between fact and fantasy we decided her musical ambitions did not have to end. We arranged public concerts that would never take place.

We walked off the bitumen streets, through a paddock of cattle on saltwater couch, to the Esplanade lined with wooden buildings and drooping streetlights not yet lit. We came to the sand where a more than a dozen tidal pools reflected the twilight arch. The sun sets quickly here and amidst the pools we stood in true twilight.

I wonder if I had hoped we would be left alone, or if that jealousy is mine – the man's rather than the boy's.

No one came on to the beach to disturb our isolation. The ocean was uninhabited but for a lonely mast-light far away.

I gave her my torch. I told her to shine it into the pools and look for the reflective eyes that would indicate the animals. You almost never found them in the tidal pools on the sand and mud and I did not hold any serious hope, only I was hoping to stretch time by putting more movements into it. A thing I knew was possible. She checked every pool on our way toward the headland where my true hope was.

We left our shoes on the sand. Our children's feet found all the footholds in the rock, and a girl of twelve gives up nothing in agility to a boy. Soon we were kneeling by a captured pool, a deep one the sea had only recently left. We did not need the torch now. Its light would not penetrate that depth of water. And anyway, all that was needed was to swirl your hand in it and if the pool held a flame bug it would light like an underwater candle.

She told me she had never seen one. If there was a flame bug there tonight I wanted it to be her find.

"You try."

She put her arm in past her elbow and stirred the water.

Two came alight. She cried with delight.

"It's a good pool," I said. "We're lucky."

"Yes. We're lucky, all right."

Though she did not know how lucky we were. It was possible to come for days on end and not see one.

"Should we look in the other pools?"

"Stay here," I said.

"Yes," she agreed. "We should stay here where we've been lucky, as long as we can."

I want to say she was beautiful then, when she spoke those transcendent words. It is impossible for a girl of so few years to be truly beautiful, yet I think she must have been as normally I would have gone checking the other pools and left this one that was certain. Instead I wanted to stay where she was pleased.

I told her how flame bugs were rarely seen together like this. How their eggs float on the foam, through the air. At first she did not believe me. I assured her it was true. This was what my father had told me. Perhaps he had been speculating or restating a myth, but nothing I have learnt since has falsified it. Our shores are protected, still the flame bugs seem to have no device for coping with even small waves – no muscular foot like a limpet, nor the ability or inclination to bind themselves into crevices like urchins. They are only ever seen at night. I do not know if they are resistant to high rock temperatures and drying out or if they die when low tide coincides with the heat of day or if instinct tells them when is safe to come close to shore.

The existence of flame bugs seems to have no practical point. Nothing in the pools ever rises to snap at them. Though if they are prey to some furtive thing, their glowing when disturbed can only aid it. They cannot be eaten by humans or used for bait and they die when put in tanks. Flame bugs seem to exist only to carry light.

"Try to catch one," I said.

"I don't want to hurt it."

I laughed, happy in my better knowledge; happier I had the opportunity to share it.

She stirred up their lights then made a grab at one that was at the far side of the pool by the time she closed her fingers. She tried again and we laughed together.

"It's nearly impossible," I said.

She sighed agreement. They could not be caught.

We sat contentedly watching them for I do not know how long. Their unpredictable movements and light meant there was no possibility of boredom. Children do not possess the accumulated pasts and anticipated futures that dwarf the present – happiness in the moment is complete happiness. Since time no longer pained us it was suspended.

There must have been a point that night when we decided it was late and we should return. I do not remember the decision. My parents were native islanders and did not care how late I came home, but her family was new to the island's customs and would be worried.

I did not wish it, but I found myself delivering her to her front gate later that night. I stayed, hidden behind a fig tree, to see her father come out and pretend to be angry when he heard her footsteps on the path. He hugged her and took her in.

I was jealous. The night might have lasted forever had we not given up on it.

I never spent another evening with the girl we called Shell. Two years passed and circumstance and my shyness meant we never became the companions we might have. Though, if at any time during those two years I had been asked to choose one of my classmates as a favourite, it would have been her. This would have surprised everyone, though not, I suspect, the girl herself. Childhood relationships may be complex and not require explanation.

Our relationship was locked in that night away from the island's inhabitants, when we found ourselves at a perfect distance from both juvenile dependency and adult sexuality, where love was unthought of, unplanned, immediate and inevitable.

She moved back to the city for tenth grade. The day before she left she came unexpectedly to my house. She told me she did not want to leave the island. She had not told anyone else. She took my hand. It was the second time we had been alone together. Then she left.

Three years later I heard she had been accepted into the city conservatorium for violin. It was two years after that, having rowed back from my launch, my mother asked me if I remembered a girl who used to live on our island and pointed to a photograph in the already-old city newspaper, to a face that was hers, though I had to look twice to be certain. My mother told me she had been killed by a man in a nightclub who had baited her drink. She possessed a beautiful future, the paper said, that had been meaninglessly cut short. Did I remember her? I cannot explain why I lied and said I did not.

I went to the beach after my mother had left me. I looked out at the ocean, at the riding light of a distant boat. I was heartbroken, though I had little right to be. I had not seen the girl in more than five years. I wondered if my love should stop now, as the pessimists would have it, since it became futile with the death of its object.

I have heard it said our souls only live after death if God remembers us. I am frightened of forgetting. This clumsy attempt to write the night of her and the flame bugs is an attempt to redeem a night in time that meant something to me, in this world where not all, and ever less, of our time has meaning. Why do I remember the feeling of that night better than its forms? I cannot be sure all I have written here is factual, though it is – in some inexplicable way – true.

I am still here on the island. I will never leave. Men still fish these waters, but they do not live on the island or build their own boats and they say there is no future in living as I do. I am not concerned with the future. I am a man who most say has done little. But I have already seen more than I understand, and lost much more than I have kept.

The flame bugs are few now. Like all beautiful things, they grow fewer as the world moves degraded through time toward its end.

I walked down the beach to the headland and climbed onto the rocks. I stirred the pool, the same pool … An unlikely flame bug rose and lit.

I spoke to the creature, to the stars, to eternity, to whatever would hear me. I asked it to remember the lost and inimitable movements of that night that time had passed by.

Should we look in the other pools?

Stay here.

Yes. We should stay here where we've been lucky, as long as we can.

Why can't I keep you?

Deep in the pool a second bug lit and rose up beside the first like a fallen tear of light. ■

Patrick Holland won the 2005 Queensland Premier's Literary Award for Best Emerging Author. His novel *The Long Road of the Junkmailer* will be published by University of Queensland Press in September 2006. His story 'The angel in the travelling show' was published in *Griffith REVIEW: Up North*, Spring 2005.

Essay:
It's life, Jim but not as we know it
Author:
Ian Lilley

Image: Pleistocene human footprints: Willandra Lakes, south-eastern Australia / © Michael Amendolia 2005 with permission of the traditional land owners.

It's life, Jim, but not as we know it

There has been a great deal of environmental change in the Australian region over the 50,000 years that people have lived here. There is going to be a lot more of it in the future, too, whether or not the specifics of current climate-change models are borne out.

About 120,000 years ago, temperatures and sea levels were like today's but then began falling. By the time people first settled the continent 50,000 years ago, sea levels were perhaps thirty metres lower than now, the climate was mild and wet and many now-dry inland lakes were full. The lower parts of the Murray–Darling system provided a particularly rich environment for people. This has been revealed in stunning detail by archaeological research at Lake Mungo in south-western NSW. These days, Mungo is a flat, parched expanse of low scrub edged by the thirty-metre high, thirty-kilometre long and mightily impressive Walls of China sand dune. When the lake was full, there was an abundance of clear water, freshwater fish including Murray Cod, mussels and crustaceans. Emu and a wide variety of marsupials including the "Tasmanian" tiger also flocked to the water. Waterfowl probably did too, although their bones have not survived. The remains of all the other creatures, as well as more enduring parts of the tools and weapons used to hunt and process them, have been found in ancient shell middens and campfires in the great sand dune, indicating the extensive use people made of the lake and its surrounds tens of millennia ago.

Life at Lake Mungo and elsewhere around Australia at the time was not to last, as environmental change forced dramatic changes. Temperatures, rainfall and sea levels plummeted about 35,000 years ago and continued to fall until the Last Glacial Maximum (LGM) 25,000 to 12,000 years ago. Average sea-surface temperatures during this period were up to four degrees lower than now in the tropics and up to nine degrees lower further south. Sea levels dropped 130 metres below today's levels as the world's water was taken up in the northern hemisphere's vast icesheets.

Despite the name, there was actually very little ice in Australia during the LGM. In Europe and North America, icesheets kilometres thick relentlessly ground down huge areas of the landscape during this period. Films such as *The Day after Tomorrow,* much less the Disney cartoon *Ice Age,* do not even begin to show what it would have been like under full glacial conditions in the northern hemisphere.

Australia's climate, on the other hand, became colder but, more importantly, it also became much more arid. The cold pushed the tree line downhill in highland areas, but the main environmental shift was a major expansion of the continent's arid/semi-arid core. Today, such environments occupy nearly three-quarters of Australia's land area but then, nearly the whole continent was desert or semi-desert.

The spreading of the arid zone was accompanied by dune building on a massive scale. This process left a legacy still visible in the form of large but stabilised and well-vegetated dunes in what are now much wetter parts of the country, including Tasmania. Reflecting the patterns of prevailing winds, the dunes are arranged in a giant swirl across the continent. From space, the swirl looks like an enormous thumbprint on the land. At the time of its creation, though, things would often have been even less pleasant than usual during that period, as dust storms increased markedly in frequency and severity.

While all this was happening, the continued lowering of sea levels created land bridges between Australia, New Guinea and Tasmania, forming a super-continent scientists call Sahul. The coastline expanded elsewhere as well, dramatically increasing the size of the continent. Large, brackish lakes formed in what are now the Gulf of Carpentaria and Bass Strait. Climate change and sea-level variation had relatively little impact in New Guinea (then northern Sahul) and adjacent archipelagos, though alpine areas expanded and upland vegetation communities ringing the highlands moved downhill. Sparsely treed grassland like that now seen around Port Moresby probably expanded in southern New Guinea.

The beginning of the end of the last glacial period was marked by a rise in temperature and sea levels about 15,000 years ago, though it remained very cold and dry for about another 3,000 years. Dune building ended and tree lines began migrating back uphill. Sea-level rise broke up Sahul, with Tasmania finally separating about 14,000 years ago and New Guinea about 7,000 years ago. Australia's land area shrank considerably. John Mulvaney and Johan Kamminga paint a graphic picture of the situation at the time:

> *Vast territories, innumerable camping places, stone quarries, burial grounds, and sacred features, were submerged; the creation of continental islands | isolated populations. In most situations, the inundation of land was barely perceptible within a human lifetime, and individuals were not displaced from their … territory. But where coastal land was flat … [sea-level rise] would have been noticeable, and even a cause for concern. On the flat Great Australian Bight and Arafura Shelf between 13,000 to 11,000 years ago, the … [sea rose at] one metre a week … Regions larger than modern tribal territories would have disappeared.*

These trying times did not end with the passing of the Ice Age, either. At present, the nature of postglacial environmental shifts remains hard to pin down in the southern hemisphere. In the northern hemisphere, there was a well-defined series of arid periods (eg the Younger Dryas 12,800-11,500 years ago), but there is no clear indication that any of them occurred in the southern hemisphere. Much the same can be said of the "Medieval Warm Period" between the ninth and thirteenth centuries AD, when Greenland was vegetated and thus, in fact, green. So too with the "Little Ice Age" between the fifteenth and late nineteenth centuries AD, when the Thames regularly froze over. We do know that sea levels around Australia stabilised at their present levels about 6,000 years ago, after a period when they actually rose up to a metre or so higher than they are now. Vegetation change continued until about 4,000 years ago, in tune with climatic changes that saw rainfall and temperatures continuing to fluctuate, sometimes above and sometimes below current levels.

In addition to these variations, probably the greatest long-term regional shift was the emergence and increasing strength and frequency of the bane of Australia's farmers today, El Niño–Southern Oscillation (ENSO) events. Becoming established in its current broad pattern between 7,000 and 5,000 years ago and with a peak between 3,000 and 1,000 years ago following an abrupt rise in event magnitude, ENSO significantly increases climatic variability and, in particular, can cause prolonged periods of severe drought throughout Oceania and especially in the west.

As the description of life at Lake Mungo shows, archaeologists have built up a solid outline of life in Australia during the last glacial, which came at the end of a geological epoch known as the Pleistocene. New data have produced a major change in perspectives on Pleistocene culture. Until recently, Ice Age life was portrayed as simple and static, in complete contrast with the more dynamic behaviour that supposedly emerged in postglacial times, known as the Holocene. Previous models painted the process as one of economic, social and political "intensification", or increased complexity, especially over the past 5,000 years. We now know that while cultural change never stops, the behaviour of modern humans, such as those who originally colonised Australia 50,000 years ago, is always complex. Indeed, such complexity is the marker of modern humanity in an anthropological/ archaeological sense. Without it, people could not have colonised the continent and nearby islands in the first place. This is evident from the absence of premodern humans such as *Homo erectus*, who made it to what at the time was the tip of the Asian mainland (now Java), but no further. This cultural complexity also facilitated people's adjustment to the extraordinary environmental changes that confronted them over the millennia following colonisation.

So what happened? People had spread across all of Sahul and the Bismarck and probably the Solomon Islands archipelagos by the time the environment began to deteriorate dramatically 35,000 years ago. What they did as the Ice Age proper descended upon them depended significantly on where they were. Three examples should be enough to highlight the sort of variation: the desert, south-west Tasmania and the islands of the Bismarck Archipelago.

In the expanding arid and semi-arid zones, people contracted towards better-watered refuges around the coastline and in upland areas such as the Pilbara and the Macdonell Ranges in the interior. As a result, many sites were abandoned for millennia during the LGM, as large swathes of country simply became too dry to live in. This contraction occurred despite evidence for what Peter Veth characterises as cultures of "profound adaptability" rather than of great uniformity and conservatism as was long thought. The Pleistocene archaeological record can be difficult to work with owing to preservation problems that have emerged over such a vast period. Often, all archaeologists have to study are stone artefacts. Yet, as research summarised by Veth and Sue O'Connor indicates, variations in the abundance of such material at different times and in different places, the arrays of tool types present and changes in the sorts of stone they are made from show clearly that Pleistocene desert people were able to "reconfigure their economies, mobility patterns, territorial ranges, information-exchange systems and technological organisation" to cope with continuous climate change. Interestingly, O'Connor and Veth found that aridity had its greatest local impact towards the end of the LGM and that many areas remained uninhabitable until the early postglacial period. In fact, they think some places never returned to pre-LGM levels of vegetation cover.

The Ice Age record studied by Richard Cosgrove and his colleagues in south-west Tasmania is as rich as the one in the desert is sparse. This is because many of the most ancient sites are in limestone caves that preserve large quantities of organic material to augment the assemblages of stone artefacts that can survive just about anywhere. Ice Age Tasmania was the only part of Sahul that even vaguely resembled glacial Europe or North America. Winter temperatures fell to minus fifteen degrees and summers were cool and brief. Much of the Tasmanian highlands were covered in permanent ice, and snow was common in the lowlands. Tree cover was sparse. Unlike the situation in the desert, where there was widespread abandonment of sites through the LGM, many of the Tasmanian caves were used right through the glacial maximum until about 13,000 years ago, when dense rainforest spread across the region, greatly restricting people's access. As Cosgrove describes it, the sites have yielded "staggering amounts of stone tools, animal bones, hand stencil art, charcoal from cooking fires and bone implements for piercing skins. It is not uncommon to find 250,000 bones and 40,000 stone tools in less than a cubic metre of soil. The major human prey animal was Bennett's Wallaby...[which] represents about seventy per cent of the species found in the caves."

Historical analogies suggested that people would have visited the caves seasonally, in summer, when the weather was not too foul. Fascinatingly, however, detailed study of wallaby teeth indicated people used the sites in winter, precisely when the weather was worst. Cosgrove explains this pattern in terms of "optimal foraging strategies" which targeted the wallabies when they were fattest and most thickly furred, and also least mobile and therefore most predictable to hunt. "It appears that the people of Ice-Age Tasmania were not mere victims of the capricious environment. They systematically planned their approach to their economy, structuring it in such a way as to take advantage of the resources available. It reflects the breadth of human behavioural flexibility and the ingenious approaches to problem-solving by Australia's early inhabitants."

So, too, do events and processes that were unfolding far to the north at the time, in the Bismarck Archipelago. Matthew Leavesley and other researchers have identified six "pulses" of activity in open-air and limestone cave sites in New Britain and New Ireland. The first pulse represents initial colonisation of an island world with an impoverished land fauna. In New Ireland, the sparse archaeological record suggests that people were organised in small, highly mobile groups that moved around to find the resources they required, though Robin Torrence's work on New Britain shows resources there were moved to people.

Environmental change linked with the LGM was limited in this tropical region. Yet, for reasons that are not clear at present, the second major pulse of activity coincided with the onset of glaciation elsewhere. This pulse saw Manus settled by at least 20,000 years ago, requiring an open-ocean journey of more than 200 kilometres, about seventy-five kilometres out of the sight of land. This makes it the longest sea journey up to this time anywhere in the world and the only known Pleistocene voyage beyond the limits of one-way island visibility. It implies capable marine craft and considerable navigational ability, even by modern standards. In other words, the colonisation of Manus entailed extensive planning and great skill in execution. At this time, people on New Ireland also began importing exotic wild mammals and stone, the latter from sources 350 kilometres away on New Britain. The second phase is followed by a gap in occupation through the LGM, which again remains mystifying in the absence of significant detectable environmental change in the region at the time. The third pulse, from 15,000 years ago, saw the re-occupation of sites throughout the region. In New Ireland, this pulse represents the greatest density of deposition of any period, suggesting regional populations had grown substantially. The final pulse occurred at about 12,000 to 10,000 years ago when all the New Ireland sites show their highest rates of cultural deposition, perhaps signalling some form of socio-economic intensification. In New Ireland, the fourth pulse was followed by the abandonment of all cave sites from about 8,000 years ago, indicating yet another change in local ways of life.

There are three compelling lessons to be drawn from these sketches of Ice Age life in Australia and its near neighbourhood. First, continual and sometimes quite remarkable environmental change is normal and natural. Second, the impacts of even very large-scale environmental shifts vary between regions. The human effects of climatic and other environmental variation thus need to be thought about at regional and indeed local scales as much as an all-encompassing global one, because what happens in region "x" will not necessarily occur in region "y", even when the two are relatively close together in global terms. Finally, and perhaps most importantly in the context of today's concerns, the very clear message from the archaeology – and the archaeology of technologically simple societies at that – is that people manage the change in their environments and life goes on. It may not be life as we know it now, or even of a sort that the ancestors of the people making the change would easily recognise. To my mind, though, that is cause for optimism, not alarm. The capacity to make what can be radical changes to a way of life that may have persisted for centuries shows how adaptable and resilient people can be.

It is critical in this context to understand that large-scale natural environmental change did not cease at the end of the last glacial. Unlike the science fiction of *Star Wars*, this sort of thing does not just occur "long, long ago and far, far away". It happens right here and it never stops. Thus various things happened throughout postglacial times that would have affected human lives to greater and lesser degrees. The onset of the ENSO phenomenon was the most important of these shifts in terms of the cultural changes that occurred as people managed – and in some cases seem to have taken great advantage of – its impact on their lives. ENSO produced a significant decline in rainfall and an increase in climatic variability across much of the continent. There appears to have been a south–north trend in the onset of this drier period, which started about 4,500 to 5,000 years ago in southern Australia but between 4,000 and 3,800 years ago in the north. Rainfall increased from 2,000 years ago. Climatic variability reduced at the same time but remains a characteristic feature of Australia's environment.

ENSO is linked with major changes in Australia and may also be implicated in profoundly important historical shifts in Australia's Pacific neighbourhood. In the case of Australia, archaeologists have long sought to understand a substantial upswing in many parts of the continent in the manufacture of regularly shaped, finely made and mostly small stone implements from about 5,000 years ago (the "mid-Holocene"). The tools include so-called "Bondi points" or "backed blades", adzes and various sorts of points (as in "spear" points, though many may never have been hafted to anything).

Peter Hiscock convincingly contends that there is a causal connection between the increase in the production of these artefacts and the onset of an ENSO-dominated climate. As Australia began to experience a drier and more variable climate from around 5,000 years ago, the distribution and availability of food and other resources would have become harder to predict reliably. Excavated evidence indicates that this change occurred when population numbers in many regions were probably increasing slightly, when at least some human groups were moving into new environments and when the effects of postglacial sea-level rise were still being felt. Hiscock believes that increased production of the small artefacts was one widespread cultural shift that was made to help people manage their changing environment. He hypothesises that these tools were one component of a tool kit that helped reduce risk in acquiring resources. This is because, in technological terms, such artefacts are reliable, versatile and easy to maintain. People using this sort of technology would have had an advantage in difficult circumstances, prompting the upswing in their manufacture. Backed artefact production dropped off markedly as rainfall increased over the past 2,000 years and people began accommodating new pressures.

Turning to the Pacific, Pleistocene people moved out as far as the end of the Solomon Islands chain in the same general period that they originally settled the expanded continent of Sahul. No one succeeded in getting any further into the Pacific – what archaeologists call Remote Oceania – for another 2,000 generations. This initial move beyond the Solomons about 3,000 years ago coincided with the striking increase in the frequency and intensity of ENSO events, which some scholars think assisted the colonisation process. Atholl Anderson has scrutinised existing explanations of the timing and geographic pattern of human dispersal in Remote Oceania. He proposes that the pattern of initial settlement now evident in securely dated archaeological sites demonstrates the arrival in the western Pacific of new maritime technology in the form of the sail. He thinks it was probably accompanied by paddles for steering as well as propulsion, both of which facilitated "the development of navigation with vessel controllability".

Despite the advance that such technology represented, it still only enabled travel downwind. Prevailing winds in the South Pacific are south-easterlies, which prevented travel further out into the Pacific. Seasonal reversals do occur and, in fact, were relied upon in historical times for inter-island trading. But they would have been a slow and unreliable way to move, especially over some of the longer distances involved in Polynesia. Moreover, there is clear archaeological evidence that the process of colonisation was episodic. Following earlier suggestions by other researchers, Anderson argues that anomalous westerlies created by episodic ENSO conditions would have provided the motive power for more rapid and reliable, but also punctuated, easterly migration. Tellingly for this scenario, while the major west-east movements coincided with ENSO conditions, movements in the opposite direction, for example from Melanesia into Central Micronesia, occurred during normal weather dominated by south-easterly trade winds. Plainly, ENSO was not all bad for people in the past.

In this connection, it is possible ENSO conditions also had something to do with postglacial events in Torres Strait. Bruno David, Ian McNiven and their colleagues have recently pushed back the dates for occupation in the western islands of the strait to about 8,000 years ago. At that time, the area was still part of the Australian mainland. The islands were abandoned after they were isolated by rising seas about 6,000 years ago, and saw only fleeting intermittent visitation. From 3,500 years ago, people's use of the islands increased significantly and probably involved renewed permanent settlement. David has proposed that this resettlement was connected with the same migration that took people into Remote Oceania, but careful consideration of the evidence to hand suggests that this is very unlikely. However, it seems likely that there is a broad causal link in the form of the intensification of ENSO conditions that coincided with both events.

Let me draw things to a close by reiterating a couple of points. First, casting an archaeological eye over tens of thousands of years of human history makes it clear that people get through – and sometimes can even prosper from – major environmental change. Second, it is also evident that global patterns of climate change and their effects on people's lives have to be thought about at regional and local levels. One particularly important issue in this connection, stressed repeatedly in the climate projections of the greenhouse-focused Intergovernmental Panel on Climate Change (IPCC), as well as in the scientific literature about past environmental variation, is the ambiguity or absence of links between the northern and southern hemispheres. This is because the two have largely separate oceanic and atmospheric circulation systems. Just because millennial-scale climate change buried Europe and Canada under three kilometres of ice does not mean it did the same thing in Australia. Virtually all of the modelling of future climate is based on northern hemisphere data, so we really have no idea what these projections mean for the southern hemisphere. As the IPCC stated in 2001: "The scarce data from the southern hemisphere suggest temperature changes in past centuries markedly different from those in the northern hemisphere, the only obvious similarity being the strong warming during the twentieth century."

It is not even that straightforward. In the IPCC report's executive summary, there is the rider: "A few areas of the globe have not warmed in recent decades, mainly over some parts of the southern hemisphere oceans and parts of Antarctica. No significant trends of Antarctic sea-ice extent are apparent since 1978, the period of reliable satellite measurements."

Whether this remains true is now in doubt. New ice core data indicate that Antarctic sea ice may have been shrinking since the 1950s. Other studies show Antarctic ice shelves are thinning, perhaps owing to as-yet undemonstrated rising sea temperatures, while others report that Antarctic glaciers have been retreating since 1945. The glaciers were advancing before that, suggesting that the current retreat may be part of a long-term natural cycle. Overall, however, a recent overview in *Science* shows that the reductions that are occurring are more or less in balance with ice thickening observable in other parts of

Antarctica. Part of the problem of interpreting results such as these is that different tests monitor different chronological scales of environmental change. This makes it hard to determine whether observable shorter-term variation is in fact just one element of a much longer-term natural pattern.

Issues of scale also apply to the melting of the Greenland ice sheet, which it is said will cause a terrifying seven-metre rise in sea levels around the world. We need to understand several things here. One is that the process is not like pouring a bucket of water into a full bathtub. The earth's crust is deformable. The extra weight of melt water will push the sea bed down, in turn forcing continental margins up. This will vary locally, affecting the amount the sea rises from place to place. Greenland as a whole will probably rise as well, without the weight of ice pressing down on it.

We need to appreciate that completely melting the ice sheet would require a rise in global average temperature of about eight degrees. Since the Industrial Revolution global average temperature has risen by less than one degree, so there is some way to go. Noticeable melting will begin with less than two degrees global rise. In any case, it would take 1,000 years for half the ice sheet to melt, and 3,000 years for all of it to go. Think about those figures for a moment. They might be just a flick of the eye in geological time, but they represent between fifty and 150 generations at a human scale. At that rate, we probably have just enough time to see what's coming and do something about it. People get unnecessarily alarmed about these sorts of projections because they confuse human and geological time scales.

To add to this uncertainty, there are also major differences *within* the hemispheres as well as unexpected correspondences between them that should be taken into account. A good way to emphasise this and its implications for the future is to think of past environmental change in Australia in the context of two global "slices" cutting through the continent. The first runs north–south from pole to pole and the other runs east–west around the world along the Tropic of Capricorn. The first, called the "PEP-II transect", takes in east Asia, including Japan, South and South-East Asia and Australasia. The Capricorn transect encompasses Australia, South Africa and South America. Like Australia in the southern hemisphere, but starkly unlike the rest of the northern hemisphere, the immense Tien Shan and Tibetan Plateau in western China remained largely unglaciated throughout the LGM because it was so dry (and so had no water to be made into ice). The technical explanation is that the PEP-II transect was a long way from major continental icesheets other than Antarctica and so, although the "global thermal signal" registered there during the last glacial, it was mediated by local circumstances. In most parts of the transect, in fact, the, global signal was overwhelmed by local rainfall patterns. This meant that, while places like Japan and southern New Zealand were glaciated, others were not.

Turning to the Capricorn transect, the distinctiveness of the Tien Shan and Tibetan Plateau in the northern hemisphere during the LGM is mirrored by the distinctiveness of South America's Andes in the southern hemisphere. Like Australia, the LGM in South Africa was cold and very arid. In South America, however, now-arid parts of the west coast were wet throughout the height of the glacial. Again, it was a matter of local rainfall regimes: the Andes were watered by summer monsoons drawing moisture from the Amazon basin. This is much the same reason that the South Island of New Zealand had glaciers when Australia did not: there was much more moisture to freeze. All of this means that, regardless of how accurate current modelling of future climate might be, its specifics are primarily concerned with the northern hemisphere – and Europe and North America in particular – and probably do not give us much idea of what might happen in the southern hemisphere in general or Australia in particular.

In his recent bestseller *Collapse* (Penguin, 2005), Jared Diamond examines a range of historical evidence concerning the success or failure of various past societies based on what he understands to have been their different approaches to environmental management. At the end of the book, he states that his "remaining cause of hope" for the future is that the modern world has "archaeologists to find out what happened in the past, and TV to tell everyone else about it so they can deal with change accordingly".

The story I have told you here may not make it to television, but who am I to argue with a figure of Diamond's stature about the value of archaeological knowledge in assessing our options for the future? Like him, I think humanity's past experience gives us good reason to be prudent. Just like the people in Australia's deserts in the last glacial maximum, we should control the things we can control so they don't worsen the impact of things we can't control. Yet, on the basis of what I know about the past, I remain optimistic that people will manage. In the end, the best advice this archaeologist has to offer comes from that great futurologist Douglas Adams: DON'T PANIC! ■

A version of this essay with references and footnotes is available online at www.griffith.edu.au/griffithreview

Ian Lilley is an archaeologist and reader in Aboriginal and Torres Strait Islander Studies at the University of Queensland, secretary of the World Archaeological Congress and past-president of the Australian Archaeological Association. His most recent book is *Archaeology of Oceania: Australia and the Pacific Islands* (Blackwell 2006).

Essay:
Time, gentlemen, please
Author:
George Seddon

Image: The Pinnacles, Nambung National Park, Western Australia / Source: Gettyimages.com

Time, gentlemen, please

Much of the geology I once learned is long forgotten, but what remains, indelibly, is an awareness of time, geological time. Much of my work was in the late Devonian, more than three hundred million years ago. That did not long seem strange: it was my familiar working time, full of detailed subdivisions (Frasnian, Famennian, etc.) which were the hours, minutes and seconds of that geological time clock.

Political time, with three- or four-year elections, is fleeting; anthropoid time, with about a twenty-five year turnover, is equally transient. Both are ephemeral events like the life of a butterfly, although that is not to deny that the lifespan of the butterfly is all-important to the butterfly, as it is to the anthropoids and other ephemerids.

One of the discrepancies in time scales that irks me lies in weather reports. Every day we read the expected daily maxima and minima, how these relate to the average for that time of the year, and the rainfall against the monthly and annual average. The weatherman is doing his job and what happens next is not his fault: he is forecasting – making educated guesses, in an uncertain environment. Many people, however, more or less expect the average, or near to it, and feel aggrieved if they don't get it. "The forecast was wrong again," they say. But a forecast can't be wrong, it is a forecast. All that can happen is that it not be fulfilled.

Accurate rainfall and temperature records, on which probabilities are established, have only been kept in Australia, or anywhere else for that matter, for a century, give or take a few decades either way, so the "averages" have very limited substance. Even in historic time, change has been the norm. Think of the Danes in Greenland in the eleventh century or thereabouts, successfully farming land that has since been blanketed in "permanent" ice and snow. In the early Middle Ages, Merry England was indeed merry, with wine grapes growing in the Thames Valley – yet by the seventeenth century, sometimes known as the Little Ice Age, the Thames was frozen solid in the winters and bonfires were lit on its surface to roast Manningtree oxen (whatever they are or were). There is a great account of this in Virginia Woolf's *Orlando.*

I remember being impressed, on my first visit to Copenhagen years ago, by the very steep pitch on the roofs of the older buildings that went back to that period, roofs clearly designed for efficient snow shedding, since the weight of accumulated snow could break the roofs of the day. Now Copenhagen, like all big Western cities, is wrapped in its own heat blanket, but even without it there is no need for snow-shedding roof lines.

In the same century, the Dutch first came to Rottnest, off the coast of south-western Western Australia, and their reports suggest that the island was much better timbered and much greener than it is today, implying a rainfall of about 1,200 millimetres rather than that of today, which is under 800 millimetres. Much of it is now an arid steppe, although climate change is not alone responsible for that. All these changes have been within historic time, which is still little more than the blink of an eye to a geologist.

In my most recent book, *The Old Country: Australian landscapes, plants and people* (Cambridge University Press, 2005), I have written about time and space (our space or place). For my Australian-born mother, the "Old Country" was Britain, as it was for most of her generation, although she had never set foot on its sacred soil. Neither had she ever been "Home", although she first left home (in central Queensland) at the age of twenty-three to be married and returned often. But Australia is "Home" and we should start behaving like it is. The latitude of Fremantle, where I live, is that of Marrakesh on the edge of the Sahara. The latitude of Hobart is that of Rome, although Rome tends to be colder in winter and hotter in summer. Their similarity in climate was remarked upon by several early English visitors, but if I mention that today, I am met with disbelief. Occasionally someone looks it up, and reports back with amazement: "My God, you're right". Australians no longer know their geography.

For my generation, Australia is also the old country. Many of our landscapes are immensely old. Last year I spent some time in Lazio, the countryside around Rome, and was reminded yet again how new is that landscape. We had lunch beside Lake Albano, a volcanic crater that erupted in mid-Pleistocene – yesterday to a geologist. It pushed up from the sea floor. Rome and all the countryside around it was under water two million years ago. The sharp contrast between an old, rich culture and a brand-new landscape, hardly yet dry from the sea, is not surprising. The fertility of Lazio underwrote and underwrites that cultural growth.

Although most Australian landscapes have undergone countless cycles of weathering, most of them have been high and dry for hundreds of millions of years. And Australia has not experienced the cataclysms of mountain building that resulted in the Rockies and the Sierras that transformed the "New World" and built the Alps in the "Old World". We don't do mountains here. Our "alps" are the tilted and eroded edge of a minor plateau near the eastern margin of the continent. Everything to the west of the Great Divide – hardly "Great" – is part of western Australia, a huge landmass that includes western New South Wales, western Queensland, all of South Australia and, of course, Western Australia. We would have done far better to follow the example of the Americas, both North and South, by having our continental divide about four times higher, running through Kalgoorlie. It wouldn't have made much difference to western Australia, but it would have made a huge difference to the east.

Massive glaciations scoured the northern half of both the northern continents, North America and Eurasia, in the Ice Age, wiping the slate clean, a new beginning for plants and man. Our last comparable ice age was in the Permian, not one million but more than 200 million years ago, when most of the western third of the continent was scoured and scored, the striations sometimes still to be seen in the ancient granites and gneisses of the Darling Plateau near sunny Perth.

Geology is a continuum so, like every other landmass, Australia has rocks of all ages from the Archaeozoic – almost the beginning of earth time – to the present, including comparatively recent volcanic activity in Victoria and the drowning of a substantial fringe of coastal land when the seas rose with the melting of Pleistocene ice in the northern hemisphere some 10,000 years ago. There are old rocks in every continent, but the Precambrian in Australia is exposed over a vast area and well studied, largely because of its mineral riches.

What is more important than the age of the rocks from a human point of view is the age of the soils, and even in geologically younger eastern Australia, the soils are mostly old and leached of nutrients.

As a political entity, Australia is, by contrast, a very young country. A few years ago, there was a rhetoric to go with it: we have the virility of the young, unlike the effete Europeans (other than, of course, the British, our immediate antecedents, whose virility is eternal). Now the tenses change. They change in the United States, whose youth is/was "one of its hoariest traditions", according to Oscar Wilde. They change in Australia, and in Britain. Old Father Thames may keep rolling along (into the mighty sea) but the North Sea is a drop in the ocean and young Father T hasn't been rolling into it for long, either. Some Australian rivers have been following much the same course for much longer, but we do not have enough words for degrees of "old". "Ancient" belongs to B-grade fiction. "Immemorial" sounds promising, but since nearly all of geological history is beyond the reach of human memory (is immemorial) it is of little help in discriminating degrees of "old".

There is a rough dirt track leading south from the road from Port Hedland to Marble Bar in north-western Australia that leads to an old mine site. Nearby, there are several outcrops of fossil stromatolites and they are up to

3.5 billion years old, among the oldest known evidence of living organisms. The stromatolites are in the Warrawoona Group, which are mostly volcanic lavas, but with some layers of sediment accumulated in shallow seas. Beneath these rocks, there is an angular unconformity or ancient erosion surface (a page missing in the local journal of events); the eroded rocks have been dated to 3.515 billion years. For erosion to take place, they had to be at the earth's surface. This is the first reliably dated evidence of a stable crust. The world was beginning to assume its present form, so we are watching the curtain go up. It must have gone up elsewhere, but this is the first record, first by more than half a billion years. Now that's old.

What was it like? Steamy. Volcanoes erupting, cooling lava, shallow seas, igneous rocks, mostly granitic, heaving away like hot thick porridge down below. A reducing atmosphere, mostly nitrogen and carbon dioxide, with traces only of oxygen. Yet there was life, including the cyanobacteria quietly building their layer-cake mushrooms of alternating sediment and organic films that we now call stromatolites.

They kept at it. Moving south and a billion years later, the stromatolites were abundant in what is now the Hamersley Basin. At some point in time they became photosynthetic, able to harness the energy of the sun – for all of us, for this was a step of far greater significance for our species than the landing on the moon. Most of our energy, including all the fossil fuels (rocket fuels included), our wood fires, our food, are directly or indirectly dependent on it.

But that is only part of the story. Oxygen was a toxic waste product for cyanobacteria, and they excreted it by combining it with the ferrous oxide dissolved in seawater, precipitated as ferric oxide. This is abundant in the banded iron formations, more than a kilometre thick, mined at Newman, Tom Price and Shay Gap, and exported in massive quantities. You may not be enthusiastic about stromatolites and the cyanobacteria, but given that they paid for the nation's Nissan Patrols and Miele electric ovens, television sets and holidays in Provence, Australians might at the very least spare them a kind thought in their prayers. Nearly one-third of Australia's export income comes from the sale of minerals from Western Australia, and those little micro-organisms were working away quietly on behalf of this export industry some 2.5 billion years ago – not, of course, only in Western Australia, but they are most spectacularly exposed in the Hamersleys.

Working for the future balance of trade was not their only positive contribution; as they locked up most of the ferrous ions in seawater, they released oxygen to the atmosphere and this began the next major phase of global evolution. An oxygenated atmosphere allowed the appearance of new kinds of organisms. Simple cells somehow acquired a nucleus, a handy little command post and communications centre. Another major step was the appearance of sexual reproduction requiring genetic recombination, not as efficient as simple division, but definitely more fun, and very useful in times of rapid environmental change, adding a second evolutionary mode to random mutation.

Then, towards the end of the Precambrian, soft-bodied, multicellular creatures made their debut. Once again, they were first found and are best displayed in Australia, but this time in the Flinders Range, in the Pound Quartzite at Wilpena. After that, the umpire blew the whistle, and the race was on: trilobites, primitive fish, early land animals and plants, the dinosaurs, the mammals, the mammoths, us – all this in a mere 500 million years.

When I was young, I read *The Timeless Land* by Eleanor Dark. Australia was the place where time stood still. But she got it wrong. In a sense, Australia was the place where the clock started ticking. Nor did it stand still during the up to 50,000 years of Aboriginal occupation, a time that saw dramatic changes in the climate, the flora and fauna, and in the boundaries of a continent that embraced New Guinea and Tasmania. The stories of Aboriginal adaptation to these changes are being recovered by archaeologists and increasingly by anthropologists who have begun to take seriously the Aboriginal explanations of their past, which tell of rising sea levels, parts of the land becoming offshore islands, volcanic eruptions and of large animals that once were hunted and then became extinct. "These explanations can be found in the rock engravings, in the paintings, in the songlines and in the Dreaming," according to Keith McConnochie in *Departures: How Australia reinvents itself* (MUP, 2002).

Most of the Pilbara was better watered 60,000 years ago than it is today, and remained so to about 25,000BP, but the world then began to cool as it moved into the last ice age. The Pilbara and the Western Desert became colder, windier and much more arid, and by 18,000BP they were emptied of people except along the coastal fringes (which extended further west than they do today) and in a few isolated refuges like the Hamersley, Petermann and MacDonnell Ranges, where water could still be found in deep ravines. For as long as 10,000 years, small communities survived in such refuges in the interior without any possibility of contact with one another. Only in the Holocene period did the climate begin to ameliorate, and by the time of European arrival on the continental fringes, the interior had all been repopulated, although thinly. "The view of Australian Aboriginal culture as an unchanging timeless culture in an unchanging timeless land is clearly an untenable representation," as McConnochie wrote. Time, gentlemen, please, and ladies, too.

Despite the immense geological age of most of Australia, there is one way in which it is physically young, and that is as an independent continent. It is, arguably, the youngest of the continents, along with Antarctica, from which it had broken entirely free as a new and independent landmass only by the Eocene, some 60 million years ago.

It is also very young latitudinally, which is to say that it has only just arrived at its present latitude and is still on the move. David Williamson once wrote a clever play called *Travelling North*, a title that could serve for our continent, and only for this one. India and Madagascar have also moved latitudinally, but the other continents have rifted apart by changing longitude – by drifting sideways, if you like. Australia is the only continent that has moved almost from pole to equator.

If the continent has come from the south, however, it has been colonised by our species from the north. There has been no invasion from the south other than a few penguins, and minimal additions from east and west in our own hemisphere. People have come from the north to a continent that has come from the south, with a physical history that is nothing like that of any of the lands from which people have come, not the Indonesian archipelago, not Britain, Ireland, southern Europe, not southern China or India or Vietnam. Its biorhythms are remote from those of any of the lands to the north, and they have been hard to learn. We live in old landscapes with limited water and soils of low fertility, yet with a rich flora that is, with a few exceptions, adapted to those conditions, as we are not. There is much to learn from it, but we have been slow learners. The exceptions are those plants that are hanging on by their toenails, those that evolved in an Australia that was further south. There are quite a few in the extreme south of Western Australia, like the tingles (*Eucalyptus brevistylis, E. jacksonii*), the red flowering gum (*Calophylla ficifolia*), and in Tasmania, a considerable list of gymnosperms, such as the Tasmanian pencil pine (*Athrotaxis cupressoides*), the King Billy pine (*A. selaginoides*), the Huon pine (*Lagarostrobos franklinii* [formerly *Dacrydium franklinii*]), the celery-top pine (*Phyllocladus aspleniifolius*), and so on.

We will lose all of these as we move north. The long-term consequences of our northward movement are that more and more of southern Australia will move into the arid and subarid zone, and although this is not of immediate concern, the effects appear to be reinforced by changing climate patterns, although their interpretation still involves many uncertainties.

One was explored recently in a fascinating program about the global warming hypothesis on BBC Channel 4. If the Gulf Stream were to fail, London would have a climate like other places at a similar latitude of around 51 degrees north, well north of the city of Quebec, for example. The Gulf

Stream is a circulating loop; there is an upwelling of cold water from the ocean deeps off the Caribbean that is warmed at the surface, travels north and moderates the climate of the Mediterranean and western Europe before it is cooled at the northern ocean surface and sinks again to complete the circulation. The hypothesis is that with a quite modest increase in global temperatures, the polar ice will melt (as it already seems to be doing, and has done before), there will be more and fresher water in the oceans, the differentials that drive the Gulf Stream will disappear, and London will have winters like New York and Boston.

This story might be of interest to Australians, because the Leeuwin current plays a critical role in southern Australia, comparable with that of the Gulf Stream. A warm current travels south from North West Cape down the west coast of the continent, turns east at Cape Leeuwin and continues across the south coast, through Bass Strait and then turns south down the east coast of Tasmania, petering out south of Hobart. (This is the reason Hobart has a climate similar to that of Rome, as noted earlier.) Off Perth, the current is five degrees warmer than the Indian Ocean further offshore, and Perth owes its generous rainfall of 860 mm to the current. Were it to fail, our rainfall would drop to that of other west coast cities at a similar latitude, like Safi, the port for inland Marrakesh in Morocco, which has 239mm. Only the west coasts of Europe and Australia have warm currents; all the other continental west coasts have cold currents: the Humboldt, Benguela and California currents.

A different concern is the consequence of the movement of the plate on which Australia is the major landmass. When India had its episode of "travelling north" after the break-up of Gondwana, it collided with and under the west-central Asian plate and pushed up the Himalayas. Our plate is now pushing under the South-East Asian plate, and we have already seen the beginnings of the consequences, the tsunamis and earthquakes of very recent history, and there will be many more. There has been some political debate under John Howard and others as to how far we should go in joining forces with South-East Asia. The truth of the matter is that we will join them literally, willy-nilly, although not in my time or yours. As we push north, we will be subducted beneath the South-East Asian plate, just as India was below west-central Asia. Darwin will slide under Singapore, and the Raffles hotel site will be perched on the top of a new peak higher than Everest. The high mountains are all new mountains. Erosion and time will eventually cut them to size, as they have erased them in our own flat old land.

That's us, then. Most of Australia has nutrient-poor soils. Most of it lies in the mid-latitudes, and it therefore has limited rainfall, and that appears to be declining. Meanwhile, the coastal cities continue to expand. Sydney has a generous natural rainfall but is, nevertheless, under intense pressure, as are Melbourne and Adelaide.

The problem lies not only with water deficiency, but also with water use.

I was told at a recent conference at which I was speaking that "of course, if you have children, you have to have a lawn. It's all very well for you." (We have no lawn, an area of prostrate grevilleas and banksias etc. and large areas of brick paving.) "Your children are grown up." "Of course you are right," I hastened to agree. "You have to have a lawn if you have children." I added: "Yet millions of children in southern Europe – Italy, Spain, Greece, also Egypt to Morocco – seem somehow able to grow up without one, and have done so for millennia."

So could we. The way we use water in our gardens is part of a more general problem with our water use nationally. We even grow rice by flood irrigation in a dry landscape. But garden use is symptomatic of national attitudes, a reflection of a general failure to comprehend our geography, our latitude, our geological history, where we lie in time and space, our place in the scheme of things. We should learn about it in school, but we don't.

Some years ago, I was listening to a local quiz show for children, who were asked: "When do trees lose their leaves?" The first contestants said "autumn" and won their point. This is what they had learnt, because this is what happens in Europe. The third contestant – they couldn't hear each other – thought for a bit, then said "winter". No score – but he had checked with his own experience and he was right. Most deciduous trees lose their leaves in Perth in late winter.

The Noongah people in south-western Western Australia have six names for the seasons. Summer is the period of dormancy. Rebirth (spring) comes with the first rains in late autumn. Our gardeners try to reverse the seasons. It makes more sense to adapt to natural rhythms, to understand the nature and history of our continent than it does to force nature to adapt to our misguided aspirations. The Noongah knew better. The twelve-year old boy knew better. Listen to them. It's time. ■

George Seddon is an Emeritus Professor of Environmental Science of the University of Melbourne and an honorary Senior Research Associate of the Department of English and Communication Studies at the University of Western Australia and the author of *The Old Country: Australian landscapes, plants and people* (Cambridge University Press, 2005).

Works cited

Beilharz, Peter (ed.), 2000, *The Bauman Reader*, Oxford, Blackwell.

McConnochie, Keith, "Desert Departures: Isolation, Innovation and Introversion in Ice-age Australia" in Xavier Pons, (ed.), 2002, *Departures: How Australia Reinvents Itself*, Melbourne University Press.

Reportage:
Resource managers, altruists or just farmers?
Reporter:
Robert Milliken

Image: John Davis with his son and grandson, Myandetta Station near Bourke, NSW / Photographer: Lorrie Graham

Resource managers, altruists or just farmers?

Like other farmers in Australia's cropping belt, David Kreig was dismayed by the way the weather seemed to be continuing to get hotter and drier. In 1995, he bought a property north of Hay, in south-western New South Wales, a region whose climatic history suggested he could expect at least one heavy wet season by 2000. "Once, we could always rely on getting three floods each decade," he says. "Ten years later, we haven't had one."

Many farmers facing Kreig's problem have been forced either to adapt to Australia's new climatic uncertainties or to walk off their land. His response is one that would have been unthinkable even a decade ago. In 2004, Kreig struck a deal with CO2 Australia, a Melbourne-based company, to plant rows of mallee trees between the wheat, barley and pea crops he grows on Ionavale, another property in the same region, near Griffith. The company is using the trees, or more strictly the carbon they absorb from the atmosphere, to sell as carbon credits to Origin Energy, an electricity supplier. Kreig receives a fee from CO2 Australia for using his land over the 30-year life of the trees' carbon-absorbing cycle, plus a small share from the sale of the carbon credits.

He sees it as more than just a commercial deal designed to buffer him against hard times. "I've always had a gut feeling that trees attract rain," he says. "There is a feeling around with climate change now that everything is getting drier. I thought that planting trees could try to address whatever is happening with climate change. It could be a small step towards benefiting our children."

Altruistic thinking of this sort about global warming has never featured strongly in the Australian bush. That is starting to change as farmers realise that climate change is emerging as yet one more challenge to their survival, along with the more familiar ones of debt–income ratios and the tyranny of world prices for their wool, meat and grain. Many are facing up to the uncomfortable fact that agriculture itself has been responsible for a considerable part of Australia's not inconsiderable contribution to global warming.

Agriculture is the third-largest contributor to greenhouse-gas emissions in Australia after coal-fired power stations and motor vehicles, according to the Australian Bureau of Agricultural and Resource Economics (ABARE). A legacy of land clearing stretching back generations comes into this calculation but so do the inefficient use of nitrogen soil fertilisers and other farming practices.

The Australian Greenhouse Office, the federal body overseeing Australia's response to climate change, gives a blunt account of Australia's record. Citing the National Greenhouse Gas Inventory, it reports that agriculture is responsible for about twenty per cent of Australia's greenhouse-gas emissions, higher than agriculture's worldwide average of seventeen per cent, and higher than in any other OECD country except New Zealand. Strangely, the office declined to elaborate on these figures which are available on its website, citing a ministerial directive not to speak publicly on climate change.

Elsewhere, David Ugalde of the Australian Greenhouse Office paints a sobering picture in an article he co-wrote last year in *Environmental Sciences*. Greenhouse-gas emissions from farming occur across about sixty per cent of Australia's landmass, mainly through the escape of carbon dioxide, nitrous oxide and methane. The worst offender is nitrous oxide, a byproduct of the nitrogen used as a prime ingredient of crop and pasture fertilisers. It has a global-warming potential more than 300 times higher than the more familiar greenhouse gas, carbon dioxide. Farms are responsible for seventy per cent of Australia's nitrous-oxide emissions. In some intensive crop and pasture systems, only one-fifth to a half of nitrogen applied to help plant growth can be accounted for, says Ugalde. "The rest goes missing … as downward percolation through the soil, as surface run-off … or as a range of gaseous products, including nitrous oxide." Only in rain-fed systems does the take-up rate by plants of nitrogen fertilisers rise to eighty-five per cent.

Then there is methane, with a global warming potential twenty-one times higher than carbon dioxide. Cattle and sheep are responsible for twelve per cent of Australia's total greenhouse-gas emissions through the methane they pass into the air after ingesting their feed. Ugalde describes this as a "very significant emission indeed".

Ugalde and his co-authors contend farming in Australia has passed through three stages over the past 200 years to reach this crucial environmental point. First came the pioneers, who believed the land was robust and resilient, and that they had little capacity to change it. "The expectations of pioneers were small. Mere survival was their motivation." Then came farmers, a breed of more intensive land users who believed they should change the land to improve it, and had the capacity to do so with tractors, bulldozers and other machines. By clearing native forests from hills, mountains, plains and valleys, farmers were able to expand Australia's cultivated land and make money from it.

The era Australian farming has reached now, say the authors, is that of the "resource manager", landowners who want a competitive return while improving their land's asset value. "Resource managers have enormous capacity to impact both positively and negatively on the environment … [They] are motivated by the need to protect their resource from environmental damage, or possibly by the need to protect the resource from regulators."

So what will Australian farming's next phase look like and how will policy-makers encourage farmers to adopt new ways to reduce greenhouse-gas emissions from farms? The next wave of farmers will be asked – and expected – to be what the authors call "altruists". "Agriculturalists will be expected to reduce greenhouse-gas emissions from farms for the benefit of all."

The deal between CO2 Australia and farmers such as Kreig could be an early sign of this new approach at work. CO2 Australia is a subsidiary of CO2 Group, an environmental services company. Formerly Revesco Group, a minerals exploration company, it changed its name and its focus in 2004 after looking at what Andrew Grant, CO2 Australia's managing director, calls the "emerging carbon economy". Despite the Howard Government's refusal to ratify the Kyoto Protocol on climate change, state and federal governments have set voluntary and mandatory targets for industries, such as power companies, to reduce greenhouse-gas emissions. The companies can either comply, pay fines or buy "carbon credits" equivalent to the volume by which they exceed their emission targets.

CO2 Australia has entered the business of selling carbon credits by creating "carbon sinks", plantings of mallee eucalyptus trees that capture carbon dioxide from the atmosphere and isolate it in their roots, stems and leaves. The wheat-sheep farming belt through western NSW and northern Victoria is ideal for growing mallee sinks. The region was covered in mallees before pioneers and farmers cleared the land, and the trees themselves are robust: they can survive bushfires and long droughts.

Kreig comes from a South Australian farming family and moved to the Hay-Griffith region of NSW twenty-eight years ago. He followed up the 2004 trial mallee plantings at Ionavale with another at West Burrabogie, a mainly grazing property near Hay. CO2 Australia paid him a fee for the portion of his property it used (typically up to ten per cent of a farm's production area, with a minimum mallee planting of fifty hectares), calculated at a rate comparable to the property's market land value. In return, CO2 Group owns the trees and sells carbon credits from them. Under the NSW Greenhouse Gas Abatement Scheme, each tonne of carbon dioxide captured by a mallee plantation represents one carbon credit certificate, which CO2 then sells to an emitter of greenhouse gases. "No physical product changes hands," says Kreig. "People in Sydney or elsewhere doing the polluting can buy credits from what we produce out here to fix their problem, without anything being touched."

There is another pay-off for Kreig. He estimates about eighty per cent of the 1,800 hectares at Ionavale had been cleared since the 1930s. Here, for the first time, was a company prepared to bear the cost of revegetating part of his place. The rows of mallees, planted in twelve metre-wide belts, provide shelter from hot winds during the cropping season and shade for his sheep at the Hay property. Even though the trees' role as carbon sinks will fade after about thirty years, the deal stipulates they must stay there for a hundred years. "We don't seem to have a ground salinity problem now," says Kreig. "But you don't know what's going on underneath. So I saw this as a way of looking after the environment. We've come full circle. It's a turn-around from what we had when we first came here."

Kreig's was among the first of CO2 Australia's trial plantings on seventeen properties. Andrew Grant of CO2 Australia says the company planted about 1.6 million trees on 1,000 hectares in 2005, and is planning to plant another 1,500 hectares in 2006. Its biggest contract so far involves selling carbon credits to Country Energy, a leading Australian power company, requiring planting about 30,000 hectares of mallee trees up to 2012. Grant says the response to the idea among farmers has been "terrific". It has come mainly from three groups: bigger corporate-type farmers, younger ones who are land rich but capital poor, and a new generation of tertiary-qualified farmers who are prepared to try a more progressive approach to land management than their forebears. "The leading twenty per cent of successful farming business people tend to feature."

Progressive it may be, but its capacity for producing a new generation of "altruists" in the Australian bush seems limited. In the end, farmers identify themselves by the produce they can successfully sell from the land. When times are good, as they are for many now, the temptation to hand over their land for the creation of carbon sinks will not be high. Despite the horror images of the 2002–03 drought in eastern Australia, farming has come through a real turning point since the drought of the early 1990s, says Brian Fisher, executive director of ABARE.

The earlier drought, hitting at a time of high interest rates and low commodity prices, was enough to ruin many farmers. Those who survived have become a more "savvy" lot, says Fisher; grain growers, in particular, have reaped the benefits of crop farming productivity rising at a rate of 3.2 per cent a year over the past fifteen years. "Switched-on farmers are talking about smart technology, having satellite-controlled, driverless tractors doing their cropping to avoid soil compacting in ten years' time. It's a different world."

Fran Rowe, a rural counsellor from Tottenham in western NSW, has seen her work change dramatically since the early 1990s. Then, she was negotiating with banks to help debt-ridden farmers leave the land with whatever meagre terms she could get them. Now, on the back of rising rural land values, farmers can walk away with up to $500,000 after selling their properties and paying off their debts. Rowe is a member of the Agriculture and Food Policy Reference Group, a body launched in Canberra in early 2006 to look at agricultural policy over the next twenty years. "It's become a leaner machine," she says. "There's much greater understanding among farmers of farming as a business." For this reason, she is sceptical of rural altruism towards climate change taking too much hold. Those farmers whom she has helped examine proposals for giving up part of their cultivated land to plant mallee carbon sinks tend to see it more as a way of raising cash to reduce their debts and to stave off selling their land altogether.

Fisher identifies a new set of problems from those that devastated the bush in the early 1990s. They flow from pressures on farmers to grow bigger and more efficient and to produce more cash flow – in other words, the problems of success. As farmers face pressures to work the land more intensively, this can only mean more pressures to address their greenhouse-gas contribution as well. Yet many farmers are increasingly resentful at their urban image as climate-change offenders, especially after the NSW government recently enacted legislation to stop broadacre land clearing. Farmers in western NSW, for instance, who want to clear native woody-weed invasions to open up more pasture lands, have been stopped from doing so. They say the woody weeds are responsible for causing an even worse environmental problem, soil erosion. And they argue that going from one extreme to another, wholesale clearing of the pioneer era to locking up land in a "time warp", is a counter-productive agenda designed to please urban greenies. A climate of rising city–country rancour does not bode well for an altruistic approach to climate change. ■

Robert Milliken is a journalist and the author of a number of books, including *On the Edge: the changing world of Australia's farmers* (Simon & Schuster Australia, 1992).

Memoir: **How green is my valley?**

Author: **Melissa Lucashenko**

Everybody talks about the weather but nobody does anything about it.
– Mark Twain

Talk of a hotter, wetter climate is all the go, but I beg of you, don't talk to me about rain. Since moving on to thirteen slushy hectares in one of Australia's wettest rural shires, I have become an expert on mud (viz. its textures, behaviours, tendencies to linger and stain), mould, fungi, mushrooms – edible and otherwise – the ailment of livestock known as greasy heel, ditto rain scald, cabin fever, Monopoly, Trivial Pursuit and Scrabble. And tinea.

City rain is like Tennyson's "useful trouble". This rain, be it spitting or pouring, means a day of sprinting from the impossibly distant car park into your dry warm office, and later back again. In the paved, bituminised, concreted city, rain is a simple but effective call for action. But rain in the verdant valley where I now live means the opposite, at least once you've moved your stock to high ground and finally trekked over the road to put the cover on the water pump, a job which you've procrastinated over for the past three days of gathering storm clouds.

These mundane tasks attended to, there's nothing else to do around these parts but to wait it out as the local creeks flood and the nearby Pacific adds its twice-daily tidal surge to the calculations. A Koori neighbour further up the valley, who lives on the high side of several streams, was once marooned in her powerless, windowless shack for eleven days. There was a brief hiatus on the fifth day, she recalls, when she could have gotten out but didn't bother to, and then it started raining again and it was too late. "It was OK," she told me implacably, "we had enough food."

Eleven days, imagine it! Further down the valley, our family record so far is three long, long soggy days of incarceration when the frogs – yes, the same ones I have protected from herbicides – sang so joyously that we couldn't hear the rain drumming on the roof, let alone each other.

Living outside the city, you inevitably learn to listen to the weather and the wildlife, at least a little. In such a wet spot, some of us grow more and more like Thoreau at Walden Pond, "self-appointed inspectors of snowstorms and rainstorms, who for many years do their duty faithfully".

You would too if you lived here. When summer rolls round, we begin to monitor the movement of ants and take advice from kookaburras as to whether the sky will cloud or clear. And when the heavens open, an idiosyncratic calculus evolves, known only to valley residents. A day and a half of torrential rain (or three days of steady soaking) will cover Phillips' creek crossing two kilometres outside town. Then, to the envy of their townie mates, the valley kids will be sent home early in the afternoon on buses that plough through the water without trouble. Some of these kids will be required to swim flooded creeks to help get the cattle in; others simply bed down in front of the TV or PlayStation, their hours of sky scanning in class amply rewarded by Huey.

Bizarrely, a tiny number of these children never learn to swim. A local cattle breeder was seen in the latest flood a fortnight ago, literally jumping up and down on the dry land, flinging his arms about in anguish as a prized bull calf was swept against submerged barbed wire by the currents. More amphibious farmers than he saved the animal in the end. In each deluge, some residents will also be lucky to escape with their lives. An ill-judged horse rescue last year left a millionaire father and his daughter strapped together with a lead rope and clinging to a fence post as the waters cascaded off Mount Chincogan and rushed over them, tearing at their designer clothes.

As the second day of hard rain dawns, the kids will be kept home from school. Cars begin to slow down on The Pocket road. Drivers squint against the spattering wet and exchange information through half-wound-down windows. "You'll get through Stock Route Road but it's too deep near town" or "I wouldn't risk it". Some always do risk it, of course, including the police, who a couple of years back up-ended their patrol car in the North Arm of the Brunswick and, to general hilarity, made the front page of the *Echo*. (Don't talk to them about rain, either.)

The information we receive from the land is tightly nuanced. When our front paddock floods (forming "Lake Charles", after our son Charlie) and is lapping the road, we know we are about to be cut off from civilisation at a crossing five kilometres away. And when the run-off from our ridge breaks the deep gutters of the back paddock and instead makes its exit to the creek via our lawn, we know Kez is trapped a kilometre or so up the road and may soon need the services of a half-tonne Clydesdale or a tractor.

As the rains continue into the second night, two-wheel-drives will slowly accumulate on the western side of Phillips', and their occupants will sit in misted cabins waiting patiently for the next Patrol or Land Cruiser to brave the current. Nobody is a stranger in a flood. Orders for milk, bread and cigarettes will be filled at the shop and cheerfully dropped back by the 4WD owners, until finally, on the third morning, no vehicles at all can get through.

Cut off, the valley reverts to prewar status. After the first day, Telstra will have failed us as automatically as … well, Telstra. And without mobile reception due to the surrounding hills, and with the landline now cut, there is only word-of-mouth or television to keep us informed and entertained. And if the storms that brought the rain have also knocked the power out, it's 1935 in Main Arm, folks. Even the merry thought of the saturated tourists in Byron who've spent hundreds of dollars a week to watch it piss with rain fails to console us.

Yet it could be – and it seems it probably will be – so much worse. We who live on Bundjalung land know that eventually the rain will stop, the mould will retreat and the mud will dry. Whatever climate change is going to mean for our kids, in the short term life for us will return to normal.

Not so for the inhabitants of Tuvalu, who can expect their entire country to be submerged sometime in the next few decades. And not so for the Inuit of the Arctic Circle, whose semi-traditional hunting lifestyle will be shattered by global warming even sooner. You can build houses or teepees using forest timber, but you can't build igloos with it. Nor can you catch fish through a hole in the ice or hunt caribou or seal in what has become a temperate forest zone. And once the ice, and the hunting practices and game animals go, what is the point of a culture built around a celebration of their subtleties and their spirit?

Indigenous cultures worldwide are set on the (once unshakable) foundations of geography and ecology. The essential values of hunter-gatherer lifestyles can be carried forward into industrial life, but only with great difficulty. It remains to be seen whether indigenous peoples in trauma – as the Inuit and others inevitably will be – can successfully adapt the clan, the traditions of egalitarianism, stoicism and intensely valued community, to life in suburbs and towns. How will the values of the hunting band serve an Inuit computer programmer, or a Cree teacher? These aren't rhetorical questions but serious issues about the survival of indigenous ways of life in a world vastly different to that which spawned our traditions.

A number of us urban Aborigines have managed it. I don't refer to those dark families who have, through no fault of their own, merely assimilated to the culture of *Big Brother*, MTV and Kmart while leaving almost all Aboriginal tradition behind. I am talking about the other blackfellas, we who are bicultural – fluent in both traditions – and who walk a tightrope between both the industrial and indigenous world, which in the urban situation are pretty much in the same geographic place. I am talking about the blackfellas for whom our culture means a great deal more than superficial "dots, dancing and didges" and who think and talk about the place of indigenous philosophy in the modern world.

> *Those Landcare gubbas are looking after the place, and there's nothing more Koori than looking after the land.*
>
> – Mrs Ellie Gilbert

I am talking about Kooris and Murris who are aware of the chasm between industrial and indigenous views of the "good life" and of what constitutes a proper society. And those who are aware also of the places where the two views meet and a philosophic settlement might be reached.

From the earliest colonial times there have been such indigenes to critique white society to itself and to their own people. These figures – Unaipon, Marika, Oodgeroo, and many other, anonymous, figures – have influenced the wider society in the process. As Kim Scott, co-author with Hazel Brown of the book *Kayang and Me* (Fremantle Arts Centre Press, 2005), argues, assimilation has cut both ways. Even in bad relationships, both parties are altered by the other; neither is ever left unchanged. As a Yolgnu elder told Pathways adolescent educators in Byron Shire: "You white men here are doing the same thing as we have always done. Exactly the same."

The egalitarian ethic celebrated by Henry Lawson and Banjo Paterson – the traditions of mateship that faithfully mimic the brotherhood of initiated Aboriginal men and the myriad skills of surviving from and enjoying and maintaining the land – were learnt by some colonial whites from Aboriginal people. And a significant fraction of this knowledge has been passed down into the general Australian culture – its indigenous roots long forgotten.

Australia is one of the most casually, unself-consciously racist countries in the developed world, but it is also blacker than it knows. Paul Theroux, writing in *The Happy Isles of Oceania* (Putnam, 1992), found white Aussies living and playing in Central Australia and was struck by their similarity to the Aborigines they had displaced:

These white Australians were doing – perhaps a bit more boisterously – what Aboriginals had always done there. Because there was always water at Glen Helen it had been a meeting place for the Aranda people … This water hole was known as Yapalpe, the home of the Giant Watersnake of Aboriginal myth, and over there where Estelle Digby was putting sunblock on her nose (and there was something about the gummy white sunblock that looked like Aboriginal body paint) the first shapeless Dreamtime beings emerged … It is perhaps oversimple to suggest that white Australians are Aboriginals in different T-shirts, but they are nearer to that than they would ever admit … After all, a bungalow is just another kind of humpy.

If you've ever done "hard yakka" (rather than simply worked up a sweat), if you've retreated to the same beach each summer for a month of barefoot fishing and swimming or have diligently looked after your own little patch of Australia, then you have walked in Aboriginal footsteps whether you know it or not.

Would that this influence had been greater. More Australians might have learned not just to love the place (as some indisputably do) but to listen to the land more seriously. Had more Aboriginal philosophers been valued rather than shot or packed off to missions, all Australians might have learned the careful and intense attention to detail that many of us in the valley are still forced to practise as a matter of course. And this close attention to detail would, almost inevitably, have translated to a widespread reverence for the natural world. More Australians of all ethnicities could have become "self-appointed inspectors of wetlands and forests and beaches" rather than once-a-year visitors to national parks, who bravely blaze their trail with cigarette butts and Coke cans.

Is it too late, or will we all be "rooned"? Tim Flannery tells us that we need to drive electric cars and choose the green option on our power bills and that this will go much of the way to ameliorating the damage we've done. Sounds sensible enough to me, and fairly easy too, once we bite the economic bullet.

When the alternative may be a melting of the Greenland ice sheet and a rise in ocean levels of seven metres, sticking to electric cars seems laughably simple. But what will likely have to come first – or at least concurrently – is some sense of where we are.

But listen! Is that rain I can hear? I have to go: the creek will be rising and there's nobody else at home to look after the place. ■

Melissa Lucashenko writes novels and essays when not tending to her farm. Her essay 'Who let the dogs out?' was published in *Griffith REVIEW 8: People Like Us*, Winter 2005.

Fiction:
Downstream
Author:
Matthew
Condon

Image: Construction of Snowy Mountains Hydro-Electric Scheme - Kosciusko Snowy River on Mount Kosciusko, from Charlotte Pass 1950 / National Archives of Australia

Downstream

Whatever lies under a stone
Lies under the stone of the world
The Green Centipede – Douglas Stewart

A month after the funeral of Wilfred Lampe's mother, and having not seen or heard from Wilfred, Mitchell the publican was delegated to drive out to the house in the valley.

The lucerne was so high from the front gate to the house that Mitchell was forced to drive through a straw-coloured tunnel, the stalks taller than the vehicle. Inside the open house, on the kitchen table and resting under the sugar canister, was a note: Back next year.

It was grief, they discussed quietly at the Buckley's. It did strange things to people. Grief could crowd out an empty house. Send you into unmapped terrain.

"He's gone fishing, probably," they said. It was their age-old euphemism for rectifying all of life's ailments – death, depression, money problems, diseased stock and even sin.

Wilfred had told his employers the Cranks nothing. It was days before they noticed his absence. It was mentioned as brief and passing news at the dinner table in the main house, in between discussions on the perennial problem of flyblown sheep and the titillating findings of the recent royal commission into the sly-grog rackets.

There were more than a few who thought Wilfred had gone, yet again, to find his missing sister, Astrid. His only sibling, and now the mother passed on. That he wouldn't stop until he'd tracked the girl down and brought her home. It did not matter that she had vanished from the Monaro almost twenty years before, and had, over that time, become a figure of myth, a firmament in the folklore of Dalgety. She had become an ageless character in their storybook. He would never stop looking for her.

So as the town theorised now about the whereabouts of Wilfred Lampe, he was fewer than sixty kilometres away, as the crow flies, leading a team of drillers and surveyors into deep, mountainous scrub near Geehi. He had become a bush guide for the hydro-electric scheme.

They were partly right, back in the Buckley's Crossing Hotel. He had viewed his mother's death as portentous, entangled somehow in the preparations and activity upstream.

And, indeed, he had gone fishing for three days after the funeral, and landed a few slender browns, which he cooked there and then, over an open fire, on the gravelly bank. There was always something satisfying about cooking and eating a catch beside the river.

At the end of the third day, as he was packing his gear, he noticed a man's black felt hat carried by swiftly on the surface of the water. He stood with creel and rods in hand and watched the hat glide past. Within seconds it had disappeared around the bend.

The next day he rode into Cooma and found the pink-cheeked Dunphy and signed up with the scheme. He didn't care about the money, though they immediately provided saddlebags and supplies and even a new pair of boots. He had to see, first-hand, what they planned to do with the river.

That night, he drank beer in a pub crowded with freshly arrived workers. Dunphy, across the bar, tipped his pork-pie hat at Wilfred.

Men were playing cards and backgammon at several tables. Wilfred's boots felt tight and ill-fitting.

Someone tapped him on the shoulder.

"Herr Wilfreed. No. *Meester* Wilfreed. I am pleasured to make your acquaintance."

Wilfred turned and faced a short, sandy-haired man in a black suit and tie. The wiry man held out his hand in greeting, his elbow seemingly pinned to his ribs.

"Mr Wilfreed. I am Boris Hintendorfer, sir. I will be your accompaniment to the camp tomorrow."

They shook firmly and once.

"Mr Dunphy is informing me that you are the expert of the Australian boosh that will take us to our camp site, sir."

"That's right." Dunphy beamed across the room, his cheeks aflame, and touched the brim of his hat again.

"Very good, Mr Wilfreed. A beer for you, yes?"

For more than an hour Hintendorfer sat with Wilfred and asked questions about the terrain and the river and creek systems. Each time the German struggled with his English his eyelids fluttered uncontrollably.

"I am accustomed to the forests around Hamburg, sir, but familiar I am not with this they call the boosh."

"You'll be familiar with it soon enough."

"Yes, Mr Wilfreed, thank you, but can you illuminate on the animals of danger that we may encounter. As team leader I am to be aware of all possibilities."

He jotted down pertinent facts in a small, black flip-top notebook, nodding and fluttering as he scribbled with the pencil.

"Ja?" he said as he wrote. "Ja?"

They drank more beer. A scuffle had broken out over at one of the backgammon tables. A dozen men rushed to placate the players.

"I had a German teacher once," Wilfred said, almost to himself. "Mr Schweigestill."

"Schweigestill. Ja. Das ist Deutsche."

"He created an opera."

"So, Mr Wilfreed. You are to say there are none of the wolves?"

Hintendorfer licked the tip of the pencil and his heavy eyelids batted like moth wings.

The following day, Wilfred met Hintendorfer and the team at Jindabyne, and they headed over the range to the proposed camp site at Windy Creek, north of the Geehi River. They were mainly surveyors and drillers and of several nationalities – Poles, Balts, two other Germans and three from Norway. Two Italians commandeered a Land Rover that followed the team.

Hintendorfer rode in the vehicle as far as the eastern foothills of the range, then saddled up and joined Wilfred.

A fresh and steady wind funnelled down from Kosciuszko.

"This is not the boosh as they call it?" Hintendorfer said.

"Not yet."

"Is beautiful in its own way," the German said, "this that is not the boosh."

They crossed the range and into the lightly wooded scrub below the snowline. Wilfred followed an old brumby track into a steep gorge. The horses' ears twitched at the grinding of the Land Rover's gears.

It was mid-afternoon before they reached the site for the proposed camp. The Land Rover was still half an hour away. It moaned like an injured animal in the distant bush.

Hintendorfer dismounted, beaming. He inspected the thick carpet of ferns at his feet and the tree canopies. "Is quiet, Mr Wilfreed, like the cathedral."

Wilfred unpacked the horse and set up his swag for the night. The grinding of the Land Rover made him clench his teeth. He was irritated after the long ride and tired of Hintendorfer's officiousness. This small man of sharp angles and precise movements. He was a shiny piece of fresh-cast metal here in the scrub.

The next day, they cleared the site and erected neat rows of tents, a mess and rudimentary lavatories within view of the camp.

Wilfred built the central fire. Hintendorfer erected a portable outdoor office for himself beneath a large gum. He sat there going through his paperwork and could have been in a building by the Hamburg docks. He brushed seeds and insects off his papers as he worked.

For the next few weeks, Wilfred led several teams of the men into the bush for surveying and retrieving soil samples. He took two of the Norwegians down to the Geehi, where they performed several tests on the river. He sat quietly and watched them on the river bank.

There were men – like the Norwegians – with boxes of instruments standing knee-deep in creeks and rivers all across the mountains. The whole place was being measured and pinched and scratched at. Reduced to numbers and figures and dashes.

Wilfred had not been able to get out of his head the news, from Hintendorfer, that huge aerial photographic maps of the region were being prepared in Sydney.

"I am told," Hintendorfer quipped, "that your big river, your Snowy, looks like the giant ... what do you call ... question mark, from the air. You know it? The question mark?" He drew a loop in the air with his index finger.

It irritated him like a leech, the thought of the photographs taken from an aeroplane. And now the teams of men scraping at the earth.

He had had a picture in his mind of the Snowy River from Omeo up to its source since he was a boy. A private map. His own personal topography. Now he was being told it was shaped like a question mark. A question mark? And it felt somewhat obscene to him, an invasion of his private thoughts, that there were men in buildings in Sydney spreading out large maps on tables and overseeing, for the first time, his river, and the valley where he lived. He felt small and powerless at the thought of the giant men studying his world.

In the evenings, around the fire after dinner, the men talked in several languages and drank rum and now and again communicated with Wilfred in broken phrases that crackled like dry wood.

Late one night, Hintendorfer checked his watch and clapped his hands twice. "Schlafen," he said. "Big day for tomorrow."

One of the Poles – Wladyslaw Drabik – drained his tin mug and stared across the fire at the camp leader. He held the cup out to the man next to him for more rum.

"Go to hell," Drabik said. The group fell silent.

Hintendorfer stared back, tensing his jaw. "*Was*?"

"It is not the war anymore," he said, nursing the cup again with two hands and returning his gaze to the fire.

The young Hintendorfer remained silent and after five minutes repaired to his tent. The two Poles muttered to each other. The talk around the fire resumed.

Later that month, Wilfred guided Drabik into the bush north-east of the base camp for some surveying work. They were gone for three days.

At night, at their makeshift camp, they struggled with and achieved a sort of dialogue. Wilfred liked Drabik. He was not uncomfortable with silence. His black hair was dusted grey at the temples. A heavy, black-blue shadow of a beard made his face look perennially soiled.

"You farmer?" he asked Wilfred at the fire on the first evening.

"Not really."

"Farmer? Crop? What crop?"

"Sheep."

"Ahh, sheeps."

"You?"

"I? Engineer."

"Engineer."

"For the Polish Army, I am engineer. Was engineer."

"Oh."

They shared a jug of rum. Drabik rolled them both cigarettes.

"Married, you?" he asked. He lit his cigarette with the end of a burning branch he pulled from the fire.

"No."

"I married. Was married."

He fell quiet again. The heavy silence of the bush dropped over them and they could both detect the encroaching dewiness in the air. They stayed anchored to the dying fire for another half-hour before retreating to their swags.

The next night they assumed the same positions around the fire and repeated the ritual of the rum and the rolled cigarettes.

"My wife I lost. In the war," he said, handing Wilfred a rollie.

"I'm sorry."

"You see the war?"

Wilfred shook his head. He picked at the eyelets of his boots. He looked up and could see the orange light playing across Drabik's lined face. His eyes were dark and moist and for long moments he wouldn't blink. It struck Wilfred that his eyes, his whole face, looked different in the night. He became a much older man, the man he would be in fifteen or twenty years, his shoulders rounded over, the cheeks sunken and defined by sharp crescents of shadow. He was like a flower that reacted to sunlight, and closed in on itself in the dark.

"We was in Lublin. We was. She was died in Majdanek."

Wilfred drew quietly on the cigarette.

"You know Majdanek? It is the concentration camp near Lublin."

Drabik ran a hand down both sides of his face. He leant forward and shifted the wood on the fire. The cigarette dangled from the corner of his mouth as he squinted at the spark rise and smoke.

"I was in Majdanek also. For four years in Majdanek, but I did not see her after the first year and someone telled me she was died."

"I'm sorry."

"I was strong. That's why I was not died in the camp." Drabik adjusted the stew pot at the side of the fire and sat back on the damp earth. He threw his butt into the flames and immediately rolled another cigarette. "What is it, the place you kill the sheep for their meat?"

"Abattoir."

"Abattoir," he said, licking the cigarette paper and looking at Wilfred. "That was Majdanek. It was just the place for the killing. Is all. Russian soldiers. And the Greeks and the French and many others more. And us Poles, too. Of course, us. I was in control of the tractor, yes? The tractor. To take the bodies to the Krembecki Woods. At the forest I was one of them who took off the bodies to make the bonfire. You know bonfire?"

He jabbed the cigarette at the fire.

"It was like the making cake. One level, layer, layer of the bodies, and the planks, and another layer, up and up and up."

He picked a speck of tobacco from his lip and stared at the wet flake on his index finger. "The light of it, of the bonfire, you could see many kilometres away from Lublin. You know?"

Drabik rubbing his hands vigorously. "Getting cold, no? Me, I like. The cold."

Wilfred looked away. He could see Drabik's skeletal face surveying equipment at the extremity of the firelight, and the wall of bush.

That night Wilfred stayed awake for a long time looking at the stars through the treetops. He looked across the fire at Drabik in his swag and could make no shape of the man in the dark. He wasn't sure if he was there or not. He made no sound in his sleep.

He pondered the aerial maps Hintendorfer had told him about. What his world actually looked like from up there. He wondered if his shack in the valley would be visible in the photographs. Or the Buckley's. Or the Bolocco

Cemetery where his mother and father rested under granite. He wondered where he was when the picture was taken from the aeroplane.

I must be in the picture, he thought. And everyone I know would be in the picture. It was his whole world, his universe, small enough to be rolled up and secured with a rubber band.

Wilfred suddenly longed to hear a sound, anything, from Drabik.

Dew had settled on his swag. He could still see a glow from the dying fire, but no warmth came from it.

He did not want to see the mountains from the air. He did not want to be able to trace the meandering route of the Snowy River. He didn't want to look at it and know that he was in there, somewhere, so small that not even a magnifying glass could find him.

That night, Wilfred wasn't sure if he slept or not. He trembled with the cold in the swag, with the enormity of everything – the scheme, the men drilling into the earth and moving mountains and redirecting entire rivers – and the thought of the map, his life, and he not even a fleck of dust in it.

What kept him awake, too, was a question that had occurred to him as Drabik talked beside the fire, but one he could not ask. He wondered what had happened to Drabik's wife. He wondered if Drabik, as the driver of the tractor, had unwittingly transported the body of his own wife to the giant flaming cairn of human beings near the forest.

He realised he was ignorant of the great questions in life, and human suffering, and the incomprehensible evils that the world produced. He was here, in the bush at night, with one man, a single stranger, who lived with an unimaginable horror, one that would not touch Wilfred in a dozen lifetimes. Then you had to calculate how many Drabiks there were now, in the mountains, holding seeds of horror in their heads as they laboured away, day after day, at a foreign landscape.

He held himself in the envelope of the swag, and in the dark he listened hard for signs of Drabik's breathing.

The instant they blew the Bailey Bridge at old Jindabyne, Wilfred, astride his horse, felt the shock wave pass through him and knew, before the shower of river water returned to earth, that the river was dead.

They had come in throngs to watch the army bring down the bridge. There were gasps and cheers when the explosives were detonated. Cars parked in rows shone in the winter light like beetles, their headlights trained to the cleared valley and the old town. Soon it would all be under water.

Wilfred wore his old tweed jacket and his only tie. He could have taken the Humber but wanted to go by horseback. He tracked the Snowy River from Dalgety as best he could, almost to the base of the new dam wall, and was surprised by the carnival-like crowds.

He could have tied up the horse and joined the locals – Straw Weston from the old pub and Polly McGregor from the store – but he needed to see this alone. He had no desire to talk to anyone.

Sitting on the grass hillock, he tried to imagine the old town submerged in the new man-made lake. Old Ted's Snowy River Café. The hotel with its worn bench on the veranda. The church.

He thought of people's living rooms and kitchens and bedrooms deep down in the murky water. The backyards, the lanes where children played, and the trees they stole fruit off, and the gates of the old cemetery, the empty shelves and counter of the general store, still there but cold and silent under the water.

There would be no fish straight away, he thought. But they'll introduce them, and when they do they'll be swimming in the spire of the church and around Straw's bar and in and out of houses where people had eaten together at a table on winter nights and rooms where they'd made love and women had given birth and children had grown up and fought and hugged and slept. It would all belong to the fishes.

He had no inclination to see the new town up on the shore of what would be the lake, with its fresh bitumen roads and gutters and electric lighting and shiny roofs and fledgling gardens. On his last trip up to the summer leases in the high country, he'd caught a glimpse of it, laid out in grids, worked over by smoke-belching bulldozers, the outline of a town in white pegs and rows of fresh soil. It looked fleshy through the trees, like a wound on an animal's hide.

And he'd seen for himself the damming up at Island's Bend.

Yet downstream nothing seemed to be affected. Wilfred went about his daily business and checked the river like you would tend to a small child for any changes, any distress or difference in behaviour. There were none he could detect.

He still fished and caught some decent trout. It was this, the unaltered rhythm of the fishing, that made him wonder if he'd been worrying about nothing all along.

One afternoon, wading in up to his knees at one of his favourite spots at Iron Mungie, he saw something he'd never before witnessed in all the years he'd fished the river. He'd only caught a single brown in two hours when a small cloud of red dragonflies appeared above the surface of the water. Out of nowhere dozens of rainbow leapt up into the blur of insects, the water boiling with their attack, the sky alive with trout. And just as quickly the dragonflies were gone and the river surface smoothed.

It was this moment that quelled his concerns. Men had burrowed into the mountains, poured untold tons of concrete, scratched out roads where there had never been roads and diverted streams and creeks and rivers that had remained undisturbed for thousands of years, maybe millions – he had seen much of it with his own eyes – and still the river could produce something that surprised him. An intense flash of life.

When they finally blew the bridge, though, and the invisible wave of air from the explosion passed over and around him and the horse, and they were left in that horrible vacuum of silence and nothingness in its wake, he knew in his soul that the river would never be the same again.

In his head he'd found it hard to reconcile the contradictions. With so much water around, how could the river be killed? That day, he watched it rise an inch an hour after the explosion. It was like the spring thaw happening in fast motion, even though it was only July. They were filling the valley with water.

The closer the hydro-electric scheme got to completion the more confused he'd become. They were told the summer grazing leases would not be affected. Nothing would change. Australia was thinking big at last. This would be one of the greatest engineering feats the world had ever seen.

They had debated it in the front bar of the Buckley's. The closer the scheme got to completion, the less the talk.

Old Alf Brindlemere's son, Jack, with the same velvety rattle in his chest as his father, couldn't see the fuss.

"We done without it all these years and we did all right," he said. It became his mantra.

"It's progress, mate."

"The way I see it," Jack said, and his companions groaned, "they just wanted to keep the war goin', with all the dynamite and stuff. Couldn't let the war go."

"How you think they gonna' dig them tunnels for the pipes and shit, with toothpicks?"

"Country thick with bloody Yanks and Eyeties, like we haven't got the blokes here to do the job. Building a bloody nation, my arse. We got hands to do it."

"I seen your new water tank, you silly old bastard. You couldn't build a billycart for your kids. Three wheels and facin' backwards, that's you."

"I built plenty of billycarts in me time."

"You're a stone's throw from the biggest source of 'lectricity in the bloody country and you still scratchin' around with candles and kerosene out at your joint."

"That'll do me."

They cheered in the bar. They'd been waiting for it. "That'll do me."

As the years went by and the Snowy kept up its temperamental seasonal flow and Wilfred went about mending fences and shooting rabbits, as he had done forever, he felt, whenever he did think about it, a little stab of pride that the river and the mountains would be sending electricity to the east coast of Australia, lighting buildings and railway stations and houses he would never see. It made the river seem bigger to him than ever before. Invincible.

And still he found himself, on some nights, getting out of bed in the dark in the early hours of the morning, padding out onto the front veranda, and cocking his ear towards the distant Snowy and its comforting roar. Just as he had placed his ear to his dead mother's open mouth the day he found her collapsed in front of the fireplace.

Sitting on the stationary horse that day, he watched the crowds mill about the old town like it was an open-air museum. Folks having one last look at their homes. Sightseers and strangers strolling through it, hands clasped behind their backs, like it was already the ruins of Pompeii.

The water was rising. So much water.

He made his way around the back of parked cars and the straggly crowd and headed for home. He heard a single car approaching from behind, and moved to the gravel verge of the road.

The car stopped about twenty metres in front of him and a woman in a straw sunhat with a scarf around her neck and wearing large black sunglasses emerged from the passenger's seat and held up her hand, motioning Wilfred to stop.

He jerked on the reins of the horse.

"Excuse me, but would you mind?"

He looked down at her as she foraged in a shoulder bag. She pulled out a large black and silver camera.

"Just one photograph?" ■

Extract from *The Trout Opera*, to be published by Random House later this year.

Matthew Condon is a Brisbane-based novelist and journalist. His essay, "Of the bomb", was published in *Griffith REVIEW 9: Up North*, Spring 2005.

Memoir:
Return to the river
Author:
Heather Kirkpatrick

Image: Brown's 2005 return to the Franklin River / Photographer: Heather Kirkpatrick

Return to the river

I steered the raft with my paddle buried deeply by the stern and we sped down the final drop of Newland's Cascades on the Franklin River amid the din and spraying water caused by the helicopter's down draught as it perched on a rock nearby. It was 1988, five years after Australia's most famous environmental campaign, which prevented the flooding of the Franklin River in Tasmania's south-west.

The helicopter carried Bob Brown, then forty-three, the first elected Tasmanian Green MP. He was wearing a collared shirt and jacket; it was clear he was not staying long. Greeting our party warmly, he asked about our river experiences, telling us he had escaped briefly from his state parliament office to do a story for the ABC.

Seventeen years later, in 2005, he sits in the back of my five-person raft clad in a black neoprene wetsuit and brightly coloured spray jacket, buoyancy vest and helmet. He has not been on the river that changed his life since 1981.

A few days into the journey, Brown surprised me when he described his response when first invited to paddle the Franklin, as we sat on a huge rock above the Coruscades rapid in the Great Ravine. "When Paul Smith [a forester from Launceston in Tasmania's north] first asked me to come down the Franklin River with him in 1976, I thought about patching all those rafts and carting things around river obstacles and I wasn't keen. He had asked a stack of other people and none of them was silly enough to come … So I agreed if he in return came for a walk in the Western Arthurs [in south-west Tasmania] … And, of course, it turned out to be the best bargain I ever made."

Brown kept coming back. Between 1976 and 1981 he made seven descents of the fast-flowing and technically challenging river, which carves its way south-west from its headwaters in central Tasmania's Cheyne Range through numerous gorges before joining the Gordon River more than a hundred kilometres downstream near Macquarie Harbour on Tasmania's west coast.

In 1978, Brown committed himself full-time to the campaign when he became director of The Wilderness Society, forsaking his profession and income as a doctor. By then, saving the river had become his obsession.

I know how he feels about the river. I first paddled the Franklin in 1987 after graduating from an outdoor education course. I was excited and curious as I ventured down "the wild river" I had heard so much about. My senses were flooded by its beauty and pristine nature. I have kept coming back.

Finding time for this journey has been years in the dreaming. Brown's life as a senator and leader of the Australian Greens does not allow much time away. "It is never easy to find twelve days out of my routine. However, a federal election had been held in November. There was no immediate state election or parliamentary sitting ... everything fell in place."

It was a chance for his partner, Paul Thomas, to see the river for the first time. "We have been together for ten years and the Franklin has been a major episode in my life and he wanted to see it. The second thing was a very strong urge to go back to a place that I loved and had such a strong acquaintance with and owed so much to," says Brown.

Thomas is a tall, bearded and gently spoken man who runs a Tibetan rug shop in Hobart, manages a sheep farm in the Huon Valley south of Hobart and is an active member of the Greens. I appreciate his physical strength and competence, and have seated him in the front of my raft. He is excited, keen to experience the place that inspired Brown's vocation.

As director of The Wilderness Society during the campaign in the early 1980s, Brown watched the Franklin River debates from the gallery in the Tasmanian parliament with increasing frustration. He believed the parliament needed people more experienced in environmental and social-justice matters. In 1982, when the bulldozers rolled into the Franklin valley, Brown considered entering parliament.

On December 12, 1982, the first boatload of protesters left Strahan on Tasmania's west coast and headed for the dam site on the Gordon River. Brown was arrested four days later and imprisoned for more than two weeks at Risdon Prison. Over the next few months, 1,272 people were arrested and 447 imprisoned. During his prison stay, Brown finally decided he would run for parliament. He was elected the day after he came out.

On this journey down the Franklin, Brown is accompanied by his media adviser Ben Oquist, and Ben's fiancée Alex Gordon, Paul's sister Anne Foale, his brother-in-law Larry McCabe, and friend Mike Dempsey. Anna Feeley and I are the guides.

Yesterday the river rose to the perfect departure level. There is enough water to cover the rocks but not too much to be running on adrenalin all day. It is as though the river is just turning it on for Bob.

At the junction of the Collingwood River with the Franklin, we stop for a morning snack on a peninsula of polished golden and brown quartzite rocks. The natural tannin in the button-grass plains above stains the water like weak tea in the shallows to almost black in the depths. Brown wanders to the edge of the shingle beach and takes a photograph. His mood is exuberant, his enthusiasm contagious.

"I can't believe it," he comments as though reading my thoughts. "That is the easiest trip down the Collingwood, because of the 0.85-metre run in the river. We always came when it was much lower and dragged across the rocks and always got tipped out at one of those rapids up there half a kilometre ... You're a much better director of currents and paddlers than we ever were," Brown says, laughing.

Travelling down the river in 2005 is different to those descents at the height of the campaign by 2,000 or more supporters. "Going from a rubber duckie to these big rafts is like going from puffing billy to an express train. It is so much easier, of course, because there are five paddles going instead of one."

In the 1970s and 1980s, Brown's single-person "duckie" was frequently punctured and had to be blown up by mouth. Pumps weren't even considered; nor were wetsuits or helmets. Cargo was a waterproof pack and a barrel of food. Double-bladed paddles, homemade from dowel and plywood, provided propulsion.

Anna and I untie our rafts as the crews return. I push off from the shingle beach and we float and paddle downstream. The leatherwood trees are in full bloom and their white petals swirl in the current. Mist hangs around the tops of the hills and a soft light overhead accentuates the textures in the forest. Cormorants skim down the water just ahead of us. The rafts slip between boulders on smooth green tongues of water and roller-coast down wave trains in between the calm pools.

We reach Descension Gorge. I know the water is up and it won't be easy to stop between the 400-metres-long rapid. Anna and I run our boats close together as backup for any rescues. It's a hoot. We punch through deep holes, filling up with water that quickly drains through the floor eyelets of our self-baling rafts.

"Forwards. Hold on," I call. We reach the final drop and ... ooh, that hole is bigger than I thought. Here we go. The boat is nice and straight. Should be right. We go deep and spring out into the welcoming calm of the Irenabyss.

"When you come into the Irenabyss it is like you've come in from a storm outside and slammed the door. Silent, peaceful and you're at home," Brown says as we sit eating our Moroccan-lamb dinner at camp that evening.

Brown named the gorge in 1978, from the Greek words meaning peace and bottomless chasm. Foam from natural plant oils forms delicate lacework patterns in the depths of the gorge below, swirling sensuously from side to side in the surface current.

"It is one of those places – and there are many of them on the Franklin – where you wish you could press a button and just give everybody five minutes sitting here on the white quartzite rocks looking back into that gorge," he says, as he gazes upstream.

Day two. The river has only dropped a couple of centimetres, which is great. We say "Hi" to a bunch of outdoor instructors from Victoria, camped on the other side. One woman's eyes are like saucers as she spots Bob. We all decide on camp sites downstream for a few days to avoid any double bookings. We'll head to Watermelon Beach tonight. That's one Bob didn't name.

We're paddling through easy grade-two and grade-three rapids under Pyramid Peak. An eagle watches us from a charred tree trunk above the forest. I decide to record from my minidisc as we are paddling. Hope I don't drown my microphone. I have three-track audio with narration from Bob, my paddle commands and the sound of white water. A log-jammed rapid is reached and Bob has a story to tell.

"I thought my number was up in that rapid once. A loose line at the back of my raft was caught around the log and the raft was being held in underneath with me ... no drainage holes like in this raft. Paul Smith inched his way back up the river with a knife ... so I could reach back and cut the line and voom, away we went."

Everyone seems comfortable and absorbed by the ever-changing scenery. We camp on Watermelon's wide sandy beach. Paul says he hasn't thought of home or work or anything back in the city. This place is like that. You become so focused on the present. It brings people back in touch with simple living, intimately connected with nature.

He shows me a garden with tiny bonsai-like plants of huon pine, sassafras, myrtle, lichens, liverworts and violets all growing together on a square foot of rotting log. He describes it as "magical" and lays his mat and sleeping bag beside it.

We sleep under the stars tonight as the dew falls gently. The cirrus clouds from the afternoon have disappeared and a clear sky is overhead.

After a bacon-and-egg breakfast, Brown and Thomas fasten original green triangular No Dam stickers they found in a recent house move to the front of each raft. The slogan came from a state referendum where "No Dams" was not an option. "You could either vote for the Gordon-below-Franklin Dam or for the Gordon-above-Olga Dam. And there was such an outrage about that, that one in three people wrote 'No Dams' on their ballot papers. So it became a motto after '81 for the Franklin ...That was the biggest informal writing in Australian history."

During the day, we explore a sculptured side canyon at Askance Creek and swim in the pool under the huge cascade at Blushrock Falls. Downstream, the Side Slip rapid marks the entrance to the Great Ravine, the inner sanctum of the Franklin. The rafts speed down a steep fast chute, bouncing off the right-hand wall. We paddle the calm straight of Inception Reach. Walls 1,000 metres high tower in from either side. We move towards the Churn, the first of four major portages. The water is still too high to safely reach an easier portage route, so we decide to carry the gear along the trail on the left.

"Everybody ready? One. Two. Three," calls Brown as we heave the raft almost vertically up the steepest section of the trail. It is interesting and somewhat surprising to see him comfortably carrying barrels, bags and rafts with the ease with which he might deliver a fiery election speech. We complete the portage in three hours.

Camp is made for two nights at the top of Coruscades, the second major rapid in the Great Ravine. The group is swimming again. Rail, hail or shine for these folk. Not me. The water is freezing. I help Anna organise dinner.

It is an incredible place to hang out for a rest day. We are surrounded by huge canyon walls, the imposing golden face of Oriel Rock with rich pockets of forest fringing the river banks. Michael spots a platypus surfacing in the deep pool of Serenity Sound. I love this place. I'm in heaven.

On the fifth day, we raft the steepest drop on the river, the Forcett, near the finale of Coruscades. It is a sneaky move to the far left side of the fall followed by a big "hold on". We line up well and the raft buries deeply and submerges for a split second. This always gets people excited. If they don't hold on it is easy to be flicked out. We pull in at the bottom and I set my video camera running on autopilot; river safety is the priority here. Here comes Anna's boat. "Oooh", and there goes Ben. I throw my safety line but he's instantly pulled back into his boat. Nice rescue.

Fully laden rafts are hauled over the left-hand side of the third major portage, Thunderush, and the bottom section runs smoothly. Sometimes rafts get held sideways on a rock here, with upstream water pouring in, requiring ropes and pulley systems to haul them off. Not this time. At the final ravine portage of the Cauldron, an easy dragging route is completed on the right.

Many parties have had to abort their Franklin expeditions. The river claimed two lives during the campaign years and a handful of others later.

Brown describes an incident from the the first full river descent made by Johnson Dean and John Hawkins. It was their third attempt in 1959, and they finally made it all the way down the Franklin. During the first two attempts their parties walked out, leaving broken canoes behind. "Dean got stuck on the left-hand side on a rock with the rest of them on the central rock in the middle there throwing him canned food in the rain. It was a very dicey situation but they managed to haul him back," Brown, who knows Dean well, says.

In 1971, Fred Koolhof's party came down on rafts made of tractor tubes. The rafts smashed at the Cauldron. Brown says: "Koolhof is said to have been saying his prayers before he jumped in with his pack on his back because he had no option in rising water ... All four of them came out at the bottom end and pieced together the smashed remnants for enough flotation to get them through the river. The whole thing [the Great Ravine] is very beautiful, very spectacular, very inspiring and so different to what it would have become had the No Dams campaign not succeeded because we would be about a hundred metres underwater here, had that happened."

On the sixth day we encounter more challenging rapids through Propsting Gorge. The second of two portages so absorbs the attention of the group that they are surprised to suddenly see Rock Island Bend. Brown is immediately nostalgic: "This is where Peter Dombrovskis came in 1979, sat on the rocks just behind me here, waited for the moment and the misty morning to take that iconic picture of Rock Island Bend, which was reprinted more than a million times during the Franklin campaign. It headed up the campaign for the 1983 election to save the Franklin, which helped change governments and brought the Hawke government in, which stopped the dam through the High Court action."

I feel the adrenalin running again as we scout and set up safety ropes for the last major rapid on the Franklin, the 400-metre-long Newland's Cascades. The river is quite low now and Anna and I entertain our crews as our rafts bump and spin off rocks, forcing us to run a drop or two backwards when we don't have time to turn them around. We are both upright at the bottom of the rapid and that is what matters. We paddle a short stretch of calm water to camp.

Newlands, with a huge long overhang about twenty metres overhead, is my favourite camp site on the river. No tarpaulins are needed as secluded ledges provide shelter, complete with views of swirling lacework patterns in the river and the music of water droplets cascading off Shower Cliff on the other side.

The final two days are spent in the broad forests of the Lower Franklin. We have finished with the serious gorges; there are no large rapids to contend with. The mood is relaxed. The river gently flows thirty-five kilometres to the Gordon River. Limestone dominates the cliffs along the waterline, punctuated by caves, arches and waterworn pockets.

We are paddling on a mirror. The reflections are intense and when the breeze picks up, light patterns from the moving ripples play on the cliff like a light show. We refresh ourselves with regular swims; the temperature reaches 30 degrees. You really appreciate days like this on the west coast where it rains about 300 days a year. I lie on my back and float in the current, watching the clouds and trees move by.

Tonight at our Snake Island camp we toast Dick Smith with mugs of Tasmanian wine. He helicoptered in the box of goodies and a note: "To Bob Brown. Thanks for saving the Franklin. This gravel bank would be fifty metres underwater but for your work. I dips me lid to you. Thanks, Dick Smith." When we arrived on the beach at Flat Island this afternoon they were there waiting for us.

Everyone sleeps under the stars. In a couple of days I will miss the sound of gentle waters running over the shingle rapids as I fall asleep. I wake at three am to a huge full moon glowing through the mist before the imposing Elliot Range. I think about getting up to film but I drift off.

The next morning we climb up from the river to the Kutikina Cave. It is the size of a large living room and the site of the southernmost known habitation of humans anywhere on earth during the last ice age. Brown recalls his memories of its historic rediscovery in 1981, on a trip he made with Kevin Kiernan and Bob Burton. From the river, he and Burton heard Kiernan's cry of delight from the cave above. "When we came up here, Kevin was bent over this hearth. There were coals still in the depression in the ground and bones of animals which had been cooked by the last Aboriginal people. We were astounded. In the silence of the morning you could see the little breeze lifting the ferns outside, you could see shadows outside and it was as if we were in somebody else's house. You could absolutely imagine an Aboriginal family coming back off the river and finding us in there, even though they left 14,000 years ago."

The cave was central to the No Dams campaign, particularly after December 1982 when south-west Tasmania became a World Heritage Area (WHA), as it was a site of recognised international cultural importance. The *WHA Act* protects sites of international cultural importance – providing a compelling argument when the battle to save the Franklin moved to the High Court in 1983.

Eight kilometres downstream, we reach the Gordon River as a sea eagle flies overhead. There is a good flow from the Gordon Dam today and even a tailwind to assist in the final seven kilometres to the jetty at Sir John Falls. The thick rainforest on these river banks is much denser than on the Franklin.

The crumbling remains of a drill site made in 1982 for the proposed dam is almost reclaimed by the forest. I look across at Bob, who smiles. He noticed it, too. He looks at home here.

This eighth river journey may be his last, his only trip down the Franklin since the future of the river became certain. "It is like reconnecting with an old friend and [finding] that he or she is just as beautiful and rewarding as ever. The great thing is the simple joy of knowing it is there and people … are going to be floating down this river for centuries to come. Coming back closes the circle, the circle between going there in the first place, then the campaign, then almost losing it, and then entering a political system which would still dam the Franklin if it wasn't for enormous public sentiment, which is absolutely prohibitive of it now. Coming back has been a real journey. It was the journey of a great environmental epic in which I took part." ■

Heather Kirkpatrick is a journalist and outdoor education instructor who lives in Tasmania when she isn't travelling.

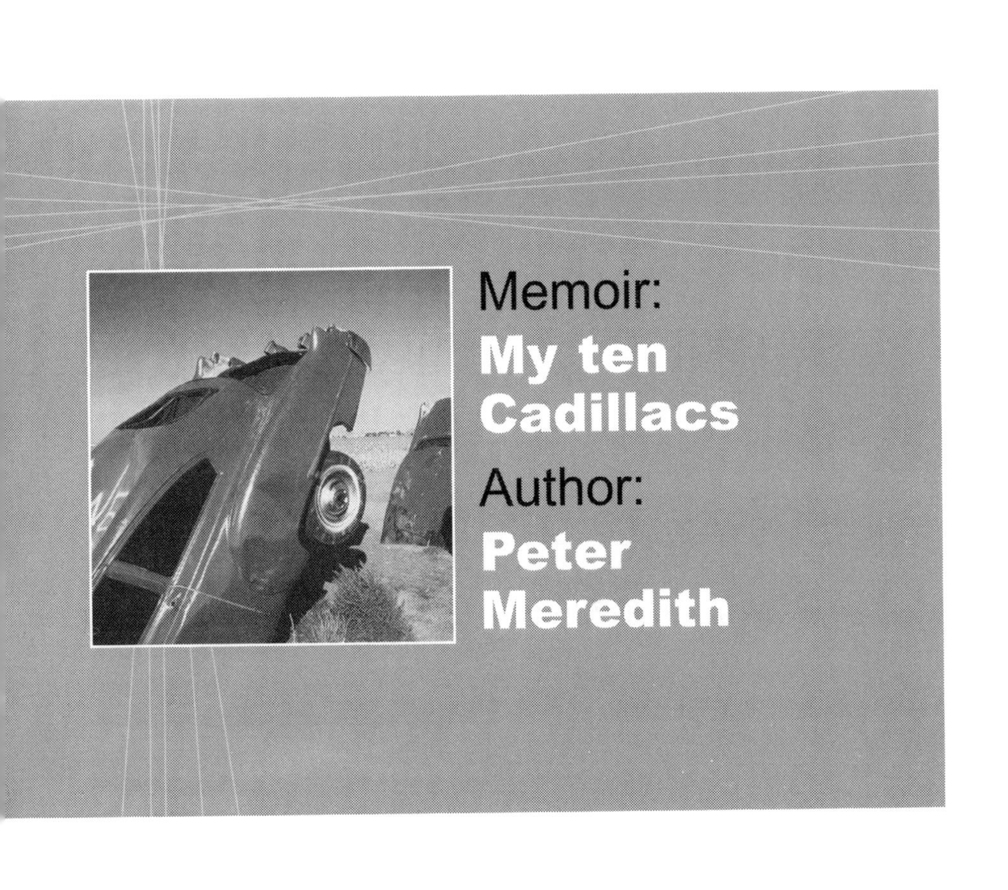
Memoir:
My ten
Cadillacs
Author:
Peter
Meredith

Image: Ant Farm / Cadillac Ranch 1974 / Courtesy of Rik Gruwez / © Rik Gruwez
Ranch owner Stanley Marsh III poses with ten Cadillacs buried as a public sculpture on his property outside Amarillo; Texas.

My ten Cadillacs

When I was nine, I won ten Cadillacs from my father in a bet. The deal was that I'd be able to claim my luxury American gas guzzlers when I was twenty-one. Until then, my father reserved the right to win them back through further bets.

The marque had a lot to do with my father's aspirations. The multinational oil company he worked for gave its employees cars precisely commensurate with their fingerhold on the slope of the corporate pyramid. At the time, my father was driving a middlebrow Ford with a radio and white-walled tyres. A Cadillac Eldorado or De Ville was a distant future prize but, given his temperament and the forces that were propelling him through life, I've no doubt he had his eye fixed on it even then.

The year was 1954. We were living in Guatemala, a small Central American country prone to earthquakes and social inequity. My father was manager of Shell's Guatemala operation, whose function was to market imported oil products around the country. The subject of our bet was the quetzal, one of the western hemisphere's showiest birds and Guatemala's national emblem. With iridescent green tail feathers up to three metres long, the quetzal lived in the remote cloud forests of Guatemala's central highlands. I'd been learning about it at school, so when my father claimed one day that it was extinct, I pounced.

"No, it's not."

"Yes, it is."

"Betcha it's not."

He looked me in the eye for a moment, sucking on his pipe and making it gurgle like the bathroom sink draining.

"Righto. I'll bet on that."

He was prone to using British idioms like that. "Righto", "righty-ho", "goodo", "cheerio", "snafu", "tally-ho, chaps" and "jolly good show", that kind of thing. He'd probably grown up with some of them in Chile before the war and picked up a good many more in the RAF during it. I was at the American School in Guatemala City and had acquired up-to-the-minute yankspeak. When he uttered those quaint expressions alone to me, I would jeer at him inwardly; when there were other people around to hear him, the sweat would break out on my upper lip.

"Okay," I said. "Twenty Cadillacs."

"Good grief! No, just one."

"Ten then."

"Righto, ten."

Although my father was an oil man, he could just as easily have been a detergent man or a home-appliances man. He would have succeeded in any business that held out to an impatient rookie the shiny prospect of a place at the top of the heap. This impatience may have had much to do with circumstance: he saw the business world as offering him an escape route from mediocrity and provincialism and he couldn't wait to hit the road.

Despite his skills and a growing bank of experience, my father insisted in later life that luck played a major, if not the major, role in his business success. He conceded that there were other people more talented and more clever than he in Shell but that he leapfrogged over them because he happened to be in the right places at the right times and knew the right people. He was, he said, the last of the uneducated to become a senior executive in the company.

Luck and contacts played their role from the start. My father joined Shell in Santiago, the Chilean capital, in 1937, at the age of fifteen. His father, Lionel Meredith, got him the job. Lionel was a Freemason; the general manager of Shell's Chilean operation was also a Freemason but junior to him. So Lionel pulled rank to give his son a leg-up.

Lionel, a stocky, good-looking fitness fanatic, had emigrated from England's midlands in 1906 to work in the nitrate industry in northern Chile. Nitrate, which yielded nitrogen for explosives, fertiliser and a range of chemicals, was Chile's biggest foreign-cash earner, although the industry was mainly British-owned. In 1910, Lionel married Maria Concepción Pinto Ceballos (always called "Conchita"), the youngest of eighteen children of a Sorbonne-educated Ecuadorian doctor and his Peruvian wife. Conchita produced a daughter and five sons, the youngest of whom was Wilfred, my father, born in 1922.

While the nitrate business flourished, Lionel and Conchita's children led charmed lives, spending their free hours horse riding in the Atacama Desert, swimming in pools and the Pacific and playing tennis. But the idyll began to crumble in 1924, when Conchita died of tuberculosis at the age of thirty-six. Then a combination of the Depression and the growing output of synthetic nitrate killed the Chilean nitrate industry. Jobless, penniless and wifeless, Lionel ended up in Santiago in about 1927 selling insurance. Life for the Merediths had become far less carefree.

Having inherited Lionel's love of physical activity, as well as his good looks, Wilfred did better at sport than academic subjects at school. The most useful skills he brought to his first job were shorthand and typing. He started as secretary, filing clerk and dogsbody to the general manager's female secretary, who took advantage of the handsome lad in the filing cupboard on more than one occasion, though not against his will.

After a while the routine became so boring that not even the hanky-panky could compensate. The only job worth having in the organisation, he decided, was the general manager's. He saw that the way to get it was to study for a qualification, gain a reputation for hard work and foster useful contacts. So he embarked on a correspondence course in automotive engineering and began to put in long hours at the office. Three years later, with a diploma in automotive engineering in his pocket, he was posted as sales rep to Antofagasta, a copper and nitrate port on Chile's mid-north coast.

World War II interrupted Wilf's upward march only briefly. Like many of his fellow expats in Chile, he volunteered to fight for Britain. Unlike most, he had a good war, flying bombers for the RAF and emerging from it unscathed, self-confident and itching to celebrate life. Having married his long-time girlfriend, Joan Irvine, another Chilean-born British expat, he was back in Antofagasta in 1950 as Shell's branch manager. By then he and Joan had three children – me (born in England in 1946) and my two younger sisters (Gill and Jean, born in Chile).

Despite his flourishing career and the nightly parties in Anto, by 1951 Wilf was beginning to feel he was slogging up a dead end. The most he could hope for after a lifelong Shell career in Chile would be the post of general manager there. He wanted more than that. He wanted to jump from the little pond into the big pond, from Shell Chile to Shell International, "to a life where horizons were broader, where I could prove myself to be as good as any other in the Shell Group".

Shell International was Big Daddy. This supranational corporation was an entity set up by Royal Dutch Shell to manage its subsidiaries around the world. Royal Dutch Shell was born in 1903 out of a merger between the Royal Dutch Petroleum Company (founded in 1890) and The "Shell" Transport and Trading Company, a United Kingdom firm (founded in 1833) that handled cooking and lamp oil, though initially it imported seashells for British collectors. In the early twentieth century, Shell expanded vigorously, buying or setting up subsidiaries in Europe, Africa and the Americas. The immediate post-World War II period marked the start of a worldwide boom in oil demand that paralleled the explosion in car sales.

For some time my father had been suggesting to Shell Chile's general manager that a trip or posting outside the country was bound to sharpen a young man's business skills. Early in 1952 his hints paid off: he was to be sent on a temporary assignment to Guatemala, though he'd still be on the Chile payroll. It was his big break, "one of those chances that hardly ever comes in a whole lifetime", as he said of it later, another of those turning points that he ascribed to luck or the hidden hand of a sponsor or patron somewhere in the corporation.

His hunger for this posting was put to the test on the morning we stepped off a disintegrating DC3 at Guatemala City's airport. The British local manager, one of the Shell dignitaries who met us, handed my father a telegram as soon as they were done with greetings. Its message was that Lionel had died in Chile while we'd been en route. To have returned to Chile for his father's funeral might have stalled Wilf's career right then. He read the telegram in silence, folded it, put it in his jacket pocket and got on with the business at hand.

We moved into the departing manager's house. Called "Las Brisas" ("The Breezes"), it had two storeys and stuccoed walls and was our biggest and most luxurious so far, though it had no phone. It came with three servants, all Mayan: a gardener with a torn ear and a grudge against the colonising race; a sad crone, the kitchen maid, who drank and had a disturbing way with a carving knife; and a sweet, plump housemaid named Marta who loved children.

My father started travelling a day or two after taking over the job. After touring Shell's facilities in Guatemala, he started on neighbouring countries. During his trips, my mother's anxiety filled the house like a fine aerosol, and the replica Roman broadsword that normally stood on the mantelpiece migrated to her bedside. She'd sit in a wicker chair on the back patio, smoking du Mauriers, darning socks or writing letters as hummingbirds drank from the honeysuckle on the balustrade. I'd sometimes find her there when I came home from school. We'd have tea and biscuits that Marta brought us on a tray, and my mother would occasionally gaze beyond the honeysuckle and the kikuyu lawn to the high wall at the bottom of the garden. A ladder was propped permanently against it. The gardener lived on the other side in a shack with his family. He'd toss bricks over the wall from his side if he knew I was nearby on our side.

When my father was home, activity in the household would move onto a higher level, as though someone had pushed a lever. The frown would lift from my mother's face and my sisters would shout and laugh and we would run about. After work my father would sometimes tell me war stories. He'd shaved off his RAF moustache by then, and though he'd filled out a little around the waist and face, to me he was still the lean, swashbuckling aeronaut who flew under a lucky star. His pipe would rattle between his half-clenched teeth as he repeated the turret gunner's warning, "Bandits at five o'clock! Bandits at five o'clock!" and his hands swooped down to show how Jerry fighters, with Spitfires on their tails, pierced his squadron of light bombers.

I daydreamed of planes and flying. We lived in the flight path of the airport and I learnt to recognise the different aircraft by the sound of their engines. At night, I'd lie awake and watch them passing over the house in my mind: a Super Constellation, a DC3, a Cessna, a Piper …

Then, at dawn on the morning of June 18, 1954, when I was eight years old, I was awoken by an aircraft I'd never heard before. It snarled in a way that made me sit up. That morning the school bus failed to arrive. My mother had the radio going in the sitting room and the maids had theirs on in the kitchen. The radios issued blasts of unfamiliar, strident music, broken by voices shouting about something called liberation. Later, two chunky fighter planes of an astonishing blue howled over our part of town and seconds afterwards there were bursts of machine-gun fire.

If my father had been there he'd have explained what was going on, I knew for sure; he'd have been able to dispel our sense of impending doom. But he was in London attending a Shell course, and my mother's mist of anxiety turned into a fog. We followed her around like a small school of minnows. She smoked and cleared her throat a lot, and when she spoke it was in a voice pitched higher than normal, her lips tightened into a little smile that was meant to reassure but didn't. This caused the maids to reply in long harangues punctuated by cries of "Ayayay!" and pleas to the Good Lord, his son and the Virgin.

She put on the same face when the torn-eared gardener came to the back patio to ask if he could move his family and animals into our garage for safety. When she said no, I worried that he might lob a lot more bricks over the wall.

The full story of what happened in the Guatemalan coup of 1954 emerged years later. Suffice to say the Americans instigated an armed putsch that ousted Guatemala's democratically elected president, whose left-wing ideas and communist sympathies the Americans considered a threat. A shambolic ground offensive by right-wing paramilitaries achieved less than some noisy but mostly harmless air attacks combined with an ingenious psychological campaign delivered by radio.

During the ten days of the coup, my mother heard nothing from or of Wilf. British diplomatic officials advised her to stay at home and await orders in the event of an evacuation. There were blackouts at night during which we would have dinner behind the big sofa by the light of a tiny candle. Lying in bed, I would hear occasional shots, some far away, some closer, and now and then a bullet flew over the house.

The first air attacks filled me with terror. It was a feeling of being stalked by people bent on killing me, and me alone. "Why me?" I wondered. What had I done to earn such detestation? When the planes came over and the bombing and shooting started, the sound was of fabric ripping, as though the sky were being torn apart from horizon to horizon. It was a solid thing that battered the wind out of me like a fist.

But soon the attacks lost their sting and a high-voltage exhilaration took over. On June 25, when the planes bombed the city's military barracks, I danced on our front balustrade as I watched the aircraft dive and the distant smoke rise.

The following day, a technician from the Shell oil depot on the city's outskirts called by. We kids pushed in close to my mother as she spoke with him at the front door. The man told her in Spanish that the planes had strafed the depot before attacking the barracks but had only punctured the tanks without setting fire to them. He showed us a machine-gun bullet recovered from the site. It seemed huge in his hand, like a copper sausage, and it was slightly bent.

On Saturday June 19, the day after the first attacks, my father had read about them in the British newspapers and phoned the Foreign Office. An official assured him there was nothing to worry about. So he decided to stay in London. His course still had another week to run but, more importantly, he had interviews with senior executives on the following Monday. On those interviews rested his career's future direction: they would dictate whether he would return to Shell Chile or join Shell International and move out into the wider world. He arrived home five days after the coup was over.

The Americans installed a puppet right-wing regime, I went back to school, my father confirmed from workers in his office that the quetzal was alive and reasonably well in the highlands, and the torn-eared gardener made peace with me by giving me a slingshot to shoot hummingbirds with. I always missed, which made him cackle oddly.

Less than six months later, Shell posted my father to Brazil as regional sales manager. He was in the big league at last.

My father's lucky star shone for the next decade. After Brazil, there were postings to Mozambique and Portugal. In each case, the house and the car got bigger and the servants more numerous. Portugal was the jewel in the Shell crown. The job came with a house as big as a hotel and a chauffeur-driven Oldsmobile like an aircraft carrier.

By the time my father took over as managing director of Shell Portugal, we kids were at boarding school in England. As it had in Mozambique, the company paid for our air fares home in the holidays. In summer especially, the days in Portugal seemed to pass in a hedonistic haze. It wasn't all beaches, pools, bars, nightclubs, dinners in ritzy restaurants and parties aboard gin palaces on a glittering sea, but there were enough of those to make me remember Portugal as Elysium and not as an ugly police state under the dictator Antonio Salazar. The real world lay far beyond the horizon.

And that's how it was for my father too. "The atmosphere in which I worked was heady; so was the socialising," he wrote later. "One could easily succumb to all the sybaritic temptations and, under the disguise of 'making connections', have a very good time. Wine was cheap and plentiful and it was considered discourteous to refuse it ... I have never before or after drunk so many Alka-Seltzers as I did in my four years in Portugal. All this was not doing my health any good, although my ego was functioning all right."

Looking back later, he saw that a kind of "folie de grandeur" had set in towards the end of his tenure in Portugal. "I took it quite naturally [sic] for my chauffeur to overtake the British Ambassador's car on the way to a reception and to put me down ahead of him. I sealed my fate in Portugal when the Co-ordinator Europe, Len, came south from The Hague for a visit/inspection. I laid on a lunch of sumptuous proportions for twelve of the most distinguished businessmen and financiers in Portugal ... It was all too good to be acceptable to Len."

Len had given my father's career an upward nudge more than once in the past. This time, though, he nudged my father in the opposite direction. Dad landed with a dull thud at Shell's head office in The Hague, Holland. There he had a job he found restrictive and a salary he considered inadequate. His house was two shoeboxes on top of one another, and his car was a frumpy Ford Zodiac he bought with borrowed money. In Portugal he'd been a big wheel; in The Hague he was a small cog in a vast machine.

It looked like Wilf's last post. His bosses, some of whom knew of his sybaritic tendencies, gave him no reason to believe otherwise. They told him they saw little future for him outside Holland and that, at the age of forty-five, he should resign himself to living and working there for a long time.

After I graduated from an English university in the middle of 1968, aged twenty-two, I hung out in London doing casual work, mostly for a house-cleaning agency called Housewives' Services. I lived in a flat with a mob of acid freaks, potheads and drunkards, all formerly respectable friends of mine. I tuned into Jimi Hendrix, turned on to a pocketful of mind-altering substances and would have dropped out had I not needed cash for food. I ranted on against Big Business, population growth and a capitalist-plutocratic economic system that fed off population growth, polluted the planet and was looking to dragoon me into its ranks to further its malevolent aims.

As the economic system's ambassador, my father was doing the dragooning – by announcing that he would cut off my rent allowance in six weeks. It would therefore be wise, he said, to find a "proper" job pretty quickly. He did try to help me with my job hunting, though. He believed with every cell of his body that business life, and especially Shell business life, was the best kind of life a man could want. I knew he'd have been proud if I'd joined Shell and followed him up the pyramid.

"I want to travel and write for a living," I told him once when he was in London on a business visit from The Hague.

"How will you do that?"

"As a journalist – a foreign correspondent or a travel writer or something."

He shook his head. "I don't know anyone in journalism. From what I hear it's extremely difficult to break into. Still, it's something to aim for, I suppose. In the meantime I'll arrange an interview for you with a personnel chappie I know in Shell. I'll pay for you to get your hair cut and your suit cleaned."

I acquiesced, had a haircut and wore a clean suit to see the personnel chappie in Shell. He didn't like what I had to offer and I hated what he had to offer. So my father arranged an interview with one of his other mates in Big Business, but that didn't work out either. Then my father suggested advertising. He knew someone in one of the London agencies, he said. I reasoned to myself that at least in advertising there'd be scope for imaginative writing: I could turn out persuasive copy during the day and write brilliant novels by night. The agency put me through an aptitude test but didn't find me apt.

With the rent-allowance cut-off date looming, my letters of application grew more eloquent. Eventually, after interviewing me more than once, Big Tobacco offered me a job as a management trainee. The interviewers had hinted at the possibility of travel to exotic places and also promised a generous cigarette allowance. As I was smoking nearly forty a day at the time, this was enticing. I accepted at once and began to wear a suit five days a week and smoke the company's brands seven days a week.

Around the time I joined Big Tobacco, my father's lucky star made its last pass across the firmament. It directed his superiors to change their minds about him and post him back to South America. He went as a director of Shell Venezuela, the most important Shell production company at the time. One of his biggest tasks was to set up a marketing organisation to control the activities of Shell in the Caribbean and parts of Central and South America. His staff called him the Tsar of the Caribbean. It was his most prestigious and best-paid posting.

I was surprised, therefore, when I later visited my parents in Caracas, the Venezuelan capital, to find them living in a house that was modest after their mansion in Portugal. Granted, the house had three storeys, but this was because it was built into a hillside. It had no a garage, just a single-space carport for my father's second-hand Chev with a blowy muffler.

What was he playing at? Was he going easy on the status symbols to save money? Or was he into reverse status, so at ease in his position that he could dispense with some of its outward trappings? They were some of the many questions I never asked him.

From his earliest days with Shell, my father had relished personal contact with far-flung company representatives. During one of my holiday visits to Caracas, he was due to fly to Lake Maracaibo, in north-western Venezuela, to inspect the company's facilities there. He invited my mother, my sister Gill and me along for the ride.

Covering some 13,000 square kilometres, Lake Maracaibo is South America's biggest lake and is thought to be the second oldest lake in the world. Long before the Spanish conquest, oil had been seeping from a hill on the lake's eastern shore, providing indigenous Indians with medicines and fuel for fires. The oil came from what is now known to be Venezuela's richest deposit, one that Shell began exploiting in 1924, thereby helping to make the country the world's fifth largest oil exporter. At the time of our visit, Venezuela was producing about 3.6 million barrels a day, seventy per cent of it in the Maracaibo region, and Shell was pumping about a million barrels a day from wells in and around the lake.

A mostly empty twenty-seater Shell plane flew us from Caracas to Lagunillas, an oil town on the lake's eastern shore. I was grumpy, withdrawn and rebellious. I reckoned I had good reason to be. My job with Big Tobacco wasn't working out and environmental issues were on my mind. They centred mainly on pollution, for which I blamed the big corporations, including my father's. So it wasn't surprising that my mood failed to be lifted by what I saw from the plane as we made our landing approach. One window showed the lake, its littoral waters forested with derricks, rigs, wellheads, balance pumps and drilling barges. The other gave a view of endless scrubland sliced by pipes and dirt roads and blotched with dark patches I took to be spills.

When we stepped out of the plane, the boiling clouds of this overcast day tried to climb into my lungs. I was deeply grateful for the air-conditioning in the purple Cadillac De Ville that awaited us. However, as the middle-ranking manager who was our driver and guide sped us through the dead-flat scrub towards the oil town, the very coolness added to the feeling of unreality that overcame me as I rode in this sumptuously furnished machine, with its synthetic paisley and fake-leather upholstery, its electric windows and inaudible engine, through a landscape being ripped apart to provide fuel for it.

We followed a triple pipeline into the town on a road sealed in crude oil mixed with sand. Balance pumps dipped and rose like giant mantises at the roadside, in churchyards, in parks and among the shanties of the unskilled workers who'd flocked to the region to find oil-related jobs. In places, pipes looped over the road. We passed a drilling rig that was installing a new well on the verge. Further on, outside the gates of a tank farm, an oil-fouled dog was licking itself clean.

Because it was strictly a business trip, the rest of us had to trail after my father as deferential men guided him around and pointed every which way. We visited one of the Shell camps, where thousands of people lived and worked in air-conditioned regimentation. We skimmed in an air-conditioned powerboat over the lake to unmanned pumping stations and gas-separating plants that hummed to one another across the water.

And, at one point during the day, we pulled up at a beach where we got out of the car to stare at black-stained rocks and a tableau of steel spires on grey water. I thought: "This is where it starts. This is where the stuff comes from that powers that Cadillac, gave Dad the life he so cherished, raised me, educated me, made me what I am and flew me halfway around the world to this spot to contemplate the meaning of it all."

I glanced about, wondering how the place looked before the Spanish came and how it would look after the oil ran out.

As we drove on, I eyed the little Cadillac logos dotted around the inside of the car. With their shield, crown and laurel wreath symbols, they denoted, I supposed, privilege, regality and success. It struck me, not for the first time that day, how apt it was that, at the pinnacle of his career, my father should be parading around in a car he once considered the acme of motoring prestige.

I always remembered the ten Cadillacs I'd won in 1954, but I had no idea what had happened to all but one of them. My father insisted I'd lost them in bets over the years since then, but I couldn't believe I'd been that stupid so many times. I knew he was pulling a fast one, just as I knew he was deliberately stalling on handing over the one he still owed me. This had been going on for four years.

Over a dinner later during that Venezuelan holiday, I asked him, "When do you think the oil will run out?"

"What oil?"

"All the oil, Venezuela's, the world's."

"It'll never run out."

"Whoa, Dad, I can't believe you just said that! You're an oil man. You should know all about it. It's made from the bodies of tiny sea creatures that are crushed and pressure-cooked until they're chemically transformed into petroleum. The process takes millions of years and we're pumping it out as fast as we can. Of course it'll run out."

My father didn't seem moved by my outburst. He went on chewing.

"Tell you what," I said. "I'll bet you it's going to run out sooner rather than later. I'll bet you the last Cadillac. If I win, I'll take cash in lieu. If you win, you owe me nothing. How's that?"

After a moment's thought he said, "Righto."

When he flew into London for a lightning business visit some weeks later, he said to me over a meal at the RAF club, "About that oil bet. I'll give you cash."

I asked him what had prompted the change of tune, but he was vague. I suspected there hadn't been a change at all. I suspected that my father, ever the petro-optimist, was just humouring me. Whatever the case, though, I got some money from him. It wasn't enough for a Cadillac but it would buy me something second-hand in reasonable nick. I was happy with that.

After my father retired from Shell at the age of fifty-five in 1977, having surfed the postwar oil wave, he settled with Joan in cosy south-east England and died twenty years later. Shell continued its transglobal march without him, growing more successful than ever, despite its record of pollution, destruction, deception and neglect in some places. Today, operating in 143 countries, it earns millions of dollars every hour. In terms of turnover, it's the fourth largest private corporation in the world; in terms of gross profits it's the second most profitable private business after ExxonMobil.

Nevertheless, just over the horizon stalks the spectre of the dwindling of the resource that made it rich. Logic dictates that oil will run out one day, whatever the optimists say, and long before then, things will get difficult for societies addicted to it. So, like most if its rivals, Shell is dipping its toe into the waters of sustainability and renewables.

In a wheat field near Amarillo, Texas, ten graffitied Cadillacs stand in a line planted nose-first in the ground, their tail fins pointing into the sky at the same angle. I sometimes fantasise that they were the ten cars I won off my father, magically translated through time and space to that spot. But the truth is that they were put there by a bunch of artists in 1974. Since then, these classic Cadillacs have become one of America's most famous public art works. Chip Lord, one of its creators, says the work can be seen as a symbol of the decline of the American empire. I see it more as portent of the end of the oil age. ■

Peter Meredith is a journalist and author with an interest in science and the environment. He lives in Bowral, NSW, and drives a 1976 Kombi van. His essay 'The ugly cousin's visit' was published in *Griffith REVIEW 10: Family Politics*, Summer 2005–06.